The Creative Brain

The Creative Brain

Myths and Truths

Anna Abraham

The MIT Press
Cambridge, Massachusetts
London, England

The MIT Press would like to thank the anonymous peer reviewers who provided comments on drafts of this book. The generous work of academic experts is essential for establishing the authority and quality of our publications. We acknowledge with gratitude the contributions of these otherwise uncredited readers.

This book was set in ITC Stone Serif Std and ITC Stone Sans Std by New Best-set Typesetters Ltd. Printed and bound in the United States of America.

Library of Congress Cataloging-in-Publication Data

Names: Abraham, Anna, 1977- author.
Title: The creative brain : myths and truths / Anna Abraham.
Description: Cambridge, Massachusetts : The MIT Press, [2024] | Includes bibliographical references and index.
Identifiers: LCCN 2023019277 (print) | LCCN 2023019278 (ebook) | ISBN 9780262548007 (paperback) | ISBN 9780262378543 (epub) | ISBN 9780262378550 (pdf)
Subjects: LCSH: Creative ability. | Intellect. | Hallucinogenic drugs— Psychological aspects.
Classification: LCC BF408 .A2347 2024 (print) | LCC BF408 (ebook) | DDC 153.3/5—dc23/eng/20231207
LC record available at https://lccn.loc.gov/2023019277
LC ebook record available at https://lccn.loc.gov/2023019278

10 9 8 7 6 5 4 3 2 1

For Andreja Bubić (1978–2021),
an extraordinary being and an even better friend.

Contents

Acknowledgments

I thank the MIT Press for the opportunity to write this book, and Philip Laughlin for his openness, kindness, and patience throughout the process.

I thank Jason Matherly in the Curriculum Library at the University of Georgia's Mary Frances Early College of Education for the many spontaneous and stimulating conversations we had that were prompted by books that I borrowed for months on end in preparation for my writing.

And I thank William Glover for the meticulous attention to detail in proofreading the final draft (I may finally have a handle on comma rules now!), and the active cheerleading from start to finish of this book project. It put the wind in my sails.

Introduction

> If a man is offered a fact which goes against his instincts, he will scrutinize it closely, and unless the evidence is overwhelming, he will refuse to believe it. If, on the other hand, he is offered something which affords a reason for acting in accordance with his instincts, he will accept it even on the slenderest evidence. The origin of myths is explained in this way.
> —Bertrand Russell (1981)[1]

There was once a well-known story about the entrancing power of the human imagination that is now either completely forgotten or so entirely unappealing that it bears no allure, even for the most superstitious among us. It began as the story of "maternal impression"—an extraordinarily prevalent notion about the human mind that was dominant in intellectual discourse from the mid-1500s until the early-nineteenth century.[2] This dated medical theory purported that the potency of maternal imagination was so singular that an over-excitement of the mind while pregnant could lead to disastrous consequences for the ensuing progeny. It was a strangely compelling idea that traced its influences as far back as ancient Greece in the works of Aristotle and Hippocrates.[3] So-called "monstrous births" included anything from bizarre physical anomalies to severe birth defects.[4]

The fact that it took several centuries for this wild idea to be fully discarded reveals much about the endurance of myths. They are incredibly hard to shake off once they have set in.[5] Even when a mountain of evidence against the established narrative accumulates, it has little persuasive sway when no (satisfying) alternative story to latch onto is offered. "The problem is that most of us find it more comforting to have certainty, even if it is premature, than to live with unsolved or unexplained mysteries."[6] Therein lies the reassurance of pseudoscience—it allows for real conviction to take root in our minds that the ideas we have about the world are grounded in reason (as opposed to superstition) because they are experienced as stable and coherent, rather than unstable and incoherent. It is not enough to know *what isn't*. One must also be able to readily grasp *what is*.

Present-day myths about the imagination are quite different in flavor. Creativity is often linked to child-like behavior and the right hemisphere of the brain, for instance.[7] Such narratives certainly seem less eccentric than the ones from bygone centuries. However, just as the myths of old, they are persuasive and impervious to change. They align seamlessly with the zeitgeist and are thereby regarded as factual by the vast majority. This makes them perfect targets for the dubious wares and catchy dogmas peddled by canny entrepreneurs and genuine do-gooders alike—dogmas that promise the unleashing of creativity by, for example, discovering your inner child or allowing for the free expression of the right brain.[8]

There are several academic works in circulation that explore the many myths that are associated with creativity with the express objective of casting them aside.[9] The aim of this book is quite different. Instead of the usual black-and-white/all-or-none tactic of debunking "myths" and endorsing "truths," I attempt a more balanced approach. In each of the myths/truths that are examined (seven in total, one per chapter), I identify their origin and how they came to be propagated. But the exploration doesn't stop there. I go further in making a case for a new premise to be considered, namely that *there is a kernel of truth that lies at the heart of each myth.*

This would, for one, explain the potency of myths and their dogged impermeability to revision or relinquishment. It makes sense that people are generally unwilling to disregard the evidence (which typically takes the form of subjective experiential knowledge) that guided the formation of their beliefs in the first place. Reason, after all, rarely trumps experience. Moreover, making the argument that there may be a kernel of truth within creativity myths permits the examination of the evidence in a more even-handed manner than is customary when advocating for a purely pro or contra stance. While such approaches are standard, they often bear the risk of emphasizing well-fitting evidence and ignoring all that is contrary—which incidentally makes for perfect fodder in the establishment of myths.

The approach adopted in this book allows for a clearer awareness of how powerful scientific and cultural myths develop and spread. In doing so, it sheds some light on how we can prime our minds to break away from orthodoxies, even if only temporarily, and glimpse the reality for what it is. Moreover, while the basis of widely held myths is often ascribed to the recipient (in terms of the psychological factors that render us vulnerable)[10] or the larger ecosystem within which information is received (including the machinations of the news media),[11] the spotlight in this book is placed squarely on the source of the misinformation—the science itself.

My hope is that the analyses and considerations presented in the book will help the reader realize that, counterintuitive though it may seem, there is a kind of poised confidence to be derived from embracing the uncertainty that goes hand-in-hand with examining any myth in as full a picture as possible. What's more, it turns out that the larger story around the brain basis of creativity—our wondrous ability to generate ideas that are novel and satisfying[12]—is also far more enthralling than any naïve myth would suggest.

Perhaps we would do well to bear in mind the words of the philosopher and mathematician Alfred North Whitehead (1861–1947) in this context. Whitehead is credited with being among the earliest to ever use the word "creativity,"[13] and as such brought the term into contemporary scientific discourse. "There are no whole truths; all truths are half-truths. It is trying to treat them as whole truths that plays the devil."[14]

1 The Creative Right Brain

To me it seems little short of miraculous that he could write out that sublime composition, the Ouverture to "Don Giovanni," while his wife read aloud to him. Indeed the statement that he did so during the night before its performance excites much doubt in my mind. For he would have had to write the part of each instrument separately, and when could the members of the Orchestra have studied and rehearsed their parts? Nevertheless it seems probable that Otto Jahn, his biographer, a writer of repute and standing, took trouble to verify the main point, which is that he had the faculty of thinking out a composition in its full detail and completeness before he set pen to paper, and that he could then write it down correctly while devising a fresh composition, or while concentrating his mind on some quite different subject. Have you met with any other such surprising manifestations of the twofold simultaneous action of the brain?
—Excerpt of a letter from Frances L. "Fannie" Hertz to Francis Galton dated November 17, 1896[1]

Dear Mrs Hertz, You always send me valuable information, and this about Mozart is perhaps the most extraordinary of all. There are plenty of instances in a faint degree of the mind working independently of the executive function of the hand, in carrying out an already determined plan, but none that I know of which is comparable in degree with that of Mozart. . . . I wish these wonderful people would submit themselves to tests and not leave the description of their performances to biographers.

—Excerpt of a letter from Francis Galton to Fannie Hertz dated November 18, 1896[2]

Let us for the sake of argument take this delightful anecdote about Mozart composing while his wife read aloud to him to be true. If we venture further and imagine that Mozart orchestrated this arrangement, it would be indicative of an uncanny metacognitive understanding on Mozart's part of the intricacies of his own unique creative process. He knew to conduct one part of his mind to be occupied by narrative and propositional content, delivered through the spoken word, which in turn allowed for the other part of his mind to focus solely on the extraordinary feat of rapid musical composition, a skill at which he was stupendously adept. What is at play here is the powerful notion of the "dual mind" or "double brain."

As a general theoretical principle, dualistic views of the mind long predate the popularization of the idea that the right brain is the seat of creative ideation. They are routinely put forward to explain a wide range of heterogeneous operations in models of perception, action, cognition, motivation, and emotion.[3] Physiologically informed conceptions of mental dualities coincided with efforts that focused on the localization of diverse psychological functions to different brain structures. This approach was brought to the fore in the early nineteenth century with the work of Franz Joseph Gall (1758–1826) and the advent of phrenology.[4] The principles of phrenology maintained that the brain was the seat of the mind, and that it was composed of distinct areas, each responsible for a specific faculty or ability. So far, so good. Truth be told, the essence of these ideas is still maintained in some form even in modern-day neuroscience, albeit with alternative variables of interest (e.g., processes instead of faculties, brain networks instead of brain areas). Where phrenology went wrong was on the insistence that the size of brain areas could be revealed by an examination of the contours and bumps of the cranium, which in turn served as an indication of how much those faculties were being engaged.

Gall naturally also noticed the near-perfect mirrored anatomical surface structure of the left and right hemispheres of the brain. This led him to propose that the two halves of the brain were also perfect symmetrical duplicates in terms of their functions or faculties, which came to be organized in that manner as a fail-safe against injury. He also maintained that the two hemispheres of the brain were not used in synchrony. Instead, each hemisphere was used singly and the functional use of one hemisphere could be substituted by the other hemisphere when the capacity of the former was exhausted. This idea that two halves of the brain were independent yet cooperatively engaged in a manner that was not necessarily synchronous or simultaneous was not shared by other leading physicians or even phrenologists of the time. Nonetheless, it laid the foundation for the central notion of the double brain and the focus on the role of the corpus callosum as the communication bridge between the two hemispheres.[5]

Functional Asymmetries of the Brain

The idea that the two hemispheres of the brain serve distinct functions involved breaking away from the notion that the surface structure of the brain—which is extremely similar across the left and right halves—dictates functional properties of the brain regions. The person who claimed credit for the idea that the brain is best characterized as a dual organ, as opposed to a single organ, was Arthur Wigan, a general medical practitioner from England, in his work *A New View of Insanity* (1844).[6] He proposed that the two hemispheres are functionally distinct and give rise to two minds with separate volitions. He averred that although one hemisphere is always superior and exerts control over the other in the healthy brain, each hemisphere is capable of processing distinct types of thought simultaneously.

The 1860s heralded the dawn of discussions surrounding "asymmetries" in the localization of function to brain structures—that is, a function may be subserved by brain region X in one hemisphere, but not by the homologous brain region X in the other hemisphere. While the importance of frontal lobe structures for the human faculty of language was first articulated by Gall in the 1920s and then championed by Jean-Baptiste Bouillaud (1796–1881), who developed the idea in detail, it was Paul Broca's (1824–1880) reports of damage to specific regions in the left frontal lobe ("the third frontal convolution") leading to speech loss that provided the smoking gun. Lesions in the region now commonly referred to as Broca's area, which refers to the posterior inferior frontal gyrus on the lateral surface of the brain's left hemisphere, was found to impair speech articulation, or "the faculty of coordinating the movements appropriate for articulate language"[7] (motor aphasia), while sparing general language faculties of speech perception and comprehension. This distinction between receptive and generative aspects of language was cemented in 1870s by the work of Carl Wernicke (1848–1905), which showed that damage to a nonfrontal region, again in the brain's left hemisphere (namely the posterior superior temporal gyrus), resulted in sensory aphasia, or a loss of speech comprehension.[8] This meant that left frontal regions were implicated in the motoric (articulatory) aspects of speech, while left temporal areas were involved in processing the sensory (auditory) features of language. Several other types of aphasia following damage to the left hemisphere regions went on to be identified. These informed the formulation of a detailed model of language function by Ludwig Lichtheim (1845–1928), which remains influential to present day.[9]

The importance of the left hemisphere was further established with the illustration of its dominant role in purposive or voluntary movement. Each hemisphere receives sensory information and controls the muscles on the contralateral side of the body (e.g., the right brain receives information from the left visual field; the left brain controls the motion of the right hand).[10] Damage to one side of the brain therefore results in a loss of function in the opposite side of the body. However, Hugo Liepmann (1863–1925) provided evidence that complicated this picture in 1900 by showing that left hemisphere lesions in the parietal lobe, particularly the supramarginal gyrus,[11] resulted in subtle motor disorders (apraxia) that feature an inability to carry out familiar movements on command. This was despite the willingness to perform the action, clear comprehension of the verbal instruction, and intact muscle control and coordination. The key discovery was that the left hemisphere was not only essential for right-sided movements, but also for intentional left-sided movements, as damage to the left hemisphere impaired movements with the right hand *and* the left hand. This pattern was not associated with right hemisphere damage, which only affected movements of the left hand. A functional left hemisphere therefore mattered more for purposive and intentional movement.

Given the indisputably singular role of language in communication, community, and culture in our species, it not difficult to see how the leap was made from the unequivocal evidence of left hemisphere involvement in language and willed action to the alleged attribution of its wider role to other functions that are seen as highly specialized or even unique to humankind, such as reasoning-based intellectual functions. Indeed, the early seeds of such claims were planted by many of the very same esteemed scholars who studied the functions of the brain. Broca, for instance, stated the following in 1877: "Man is, of all the animals, the one whose brain in the normal state is most asymmetrical. He is also the one who possesses the most acquired faculties. Among these faculties . . . the faculty of articulate language holds pride of place. It is this that distinguishes us most clearly from the animals."[12] The sought-after attributes in dichotomies (humanness, volition, intelligence, consciousness, reason) soon began to be ascribed to the left hemisphere, with the less desirable alternatives (animality, instinct, passion/emotion, unconsciousness, madness) consigned to the right hemisphere.[13] As functions of the right hemisphere received very little direct focus at the time, many of these ascribed functions of the right hemisphere were assumed or overgeneralized from limited evidence, and took the form of counterparts of what was known or alleged in relation to the left hemisphere. "Thus, it began to be argued that the right hemisphere played a predominant role in passive sensibility, emotion, activities serving trophic, instinctual life, sleep, unconscious thought processes, criminality, and madness—in this sense, neatly complementing the assertive, civilized, intellectual activities of the left hemisphere."[14] The stage for one of the most powerful memes of all was thus set: the civilized left brain versus the untamed right brain.

The Left Brain versus the Right Brain

The question of how the two functionally asymmetric halves of the brain work together led to interesting early postulations about the flexibility of the communication between the two halves. On the one hand, evidence suggested that patterns of communication were variable and context dependent, at least in relation to movements, and could be classified as "single" (one could replace the other), "conjoint and correspondent" (similar toward a common end), and "conjoint and different" (dissimilar toward a common end).[15] Contemporary work, on the other hand, demonstrates that each hemisphere is associated with unique functional activity patterns. Left hemisphere regions preponderantly communicate with other regions within the same hemisphere, whereas right hemisphere regions strongly interact with regions of both hemispheres.[16]

The brain structures that facilitate communication between the hemispheres are called commissures. Generally speaking, commissures refer to the neural fiber tracts that connect the left and right sides of the central nervous system and cross over the midline of the brain. The largest of the commissures in the human forebrain is the corpus callosum, comprising more than 90 percent of these tracts. Apart from the primary sensory regions (visual, auditory, and somatosensory), most of the cerebral cortex (the outer 2–3mm of the surface of the hemispheres) constitutes association cortex, regions that send and receive information via callosal fibers. In fact, the only association cortex that does not do so are the temporal poles, which are connected via another pathway, the anterior commissure.[17] It is little wonder that the corpus callosum, the most readily visible of all the commissures, received an especial focus in studies of hemispheric lateralization as the "highest

integrating organ of the brain."[18] Indeed, split-brain studies
on animals beginning in the 1950s attested to the importance
of this structure for integration and transfer of sensory-motor
and learning-memory functions between hemispheres. In the
1960s, the focus shifted to studies on split-brain human participants. The name reflects the fact that these people were
treated for the alleviation of intractable epileptic seizures
by means of a corpus callosotomy, a surgical procedure that
involved cutting through the corpus callosum, and thereby
severing the major connection between the left and right
hemispheres. There were two things that were striking about
split-brain patients. On the one hand, normal behavior,
unchanged personality, and unimpaired intellectual functioning continued, despite the drastic neurosurgical intervention.
On the other hand, the loss of inter-hemispheric communication and integration led to behavioral patterns that suggested
not one but two brains at work (for instance, being able to
effortlessly draw two different geometrical shapes with each
hand simultaneously).

The enormously influential work of Roger Sperry and his
colleagues laid bare the highly specific functional quirks typical
of each hemisphere upon behavioral testing. The left "major"
hemisphere was found to be central for the conscious perception of written language and the production of spoken and
written communication. Although the right "minor" hemisphere displayed the capacity for language comprehension,[19] it
was largely nonverbal and showed dominance for drawing and
manipulating visuospatial relations and was held "to possess
distinctively human emotional sensitivity and expression."[20]

The argument that split-brain evidence speaks to the separate perceptions, understandings, minds, and consciousnesses
for each hemisphere did not go unchallenged,[21] not least

because these insights were gleaned from a tiny neurologically atypical sample with heterogeneous types of callosal discon‑ nection (partial versus full; often involving the severing of other commissures as well). Nonetheless, the impact of these findings was monumental and lasting, both inside and outside of academia. They heralded the advocacy of blunt generaliza‑ tions about what each hemisphere's specialized roles are and how they could be (or even needed to be) enhanced or har‑ nessed for optimal functioning. It seemed not to matter that the left and right halves of the brains were not split but funda‑ mentally physically connected in the vast majority of human beings. Appeals for caution came from reputed sources,[22] but largely went unheeded. Split-brain studies captured the imagi‑ nation of the general public and by the 1970s "dichotoma‑ nia" was in full frenzy, exerting considerable influence in the applied domains of business, education, mental health, New Age philosophies, social activism, and the self-help market.[23]

In the world of academic research, it paved the way for the study of hemispheric lateralization in humans and nonhuman animals.[24] It is now established that laterality is a ubiquitous principle of nervous system organization in vertebrates across perception, motor, vocalization, and emotion systems.[25] Cere‑ bral asymmetries are a cross-species phenomenon that come about through genetic and nongenetic factors.[26] Among adult humans, structural neural asymmetries (physical features of the brain) are widespread.[27] While the evidence from early brain development studies on humans for structural asymme‑ tries are unclear, the case for functional asymmetries (neuronal signaling in brain regions in relation to specific cognition and behavior) is straightforward for speech (left hemisphere domi‑ nance) relative to other nonlinguistic functional domains (right hemisphere dominance).[28]

Right hemisphere functions also received renewed focus with the dawn of the split-brain era. Damage to regions of the right hemisphere (inferior parietal as well as other cortical and subcortical areas) rather than the left hemisphere is more dominantly associated with spatial neglect, or the failure to detect, respond to, or orient to stimuli presented in the left visual field.[29] The fact that damage in right frontoparietal networks leads to spatial asymmetries even in blind people[30] points to the importance of this hemisphere in supporting attentional functions.[31] With evidence for its dominance in emotion-relevant processing as well,[32] the right hemisphere came to be regarded as the mediating site of the types of psychopathologies that are classified as disorders of awareness, social cognition, and emotional functioning.[33]

The Birth of the Creative Right Brain Idea

The seeds of this notion were first planted by Joseph Bogen (1926–2005), who together with Philip Vogel formed the neurosurgeon team who brought split-brain patients zero and one to Roger Sperry's attention. It was in fact on Bogen's request that behavioral tests were specially designed to assess lateralized functions, making him a core member of the original Sperry split-brain research group.[34]

In 1969, Bogen published a series of essays titled "The Other Side of the Brain," in which he made clear that he took issue with the characterization of the right brain as the minor organ with unspecific functions in comparison to the dominant major left hemisphere. He opposed this bias against the right hemisphere, stating that "the informational capacity for one is just as great as the other."[35] His central proposal was that unlike the "propositional" left hemisphere, the right hemisphere was "appositional" and operated in a nonlinear, synthetic mode typical of musical and artistic expression as well as dream states.[36] He went on to conjecture that the tendency to see dualities in all facets of human experience was rooted in a similar duality in the mind, a realization to which people from Eastern and tribal cultures were naturally attuned, and that the left and right hemispheres are "functionally independent and cognitively complementary" in all individuals.[37] Bogen also made the rather nuanced distinction between technical proficiency in the arts (left hemisphere function) and intuitive talent in the arts (right hemisphere function), and averred that the left hemisphere exerted an inhibitory influence on the right hemisphere in the context of creativity and intuition.[38]

Unlike the assertions about left hemisphere functions that were rooted in behavioral data from split-brain studies, Bogen's assertions about the right hemisphere were not based on much in the way of direct empirical evidence. Nonetheless, the fact that these ideas were advocated for by the right source (through scientific publications by a Sperry team member) at the right time (considering the significant sociocultural upheavals experienced in the United States over the 1960s and 1970s) was the catalyst that ignited the powerful creative right brain meme.[39] Any finding that appeared to fit this pattern, such as higher left-handedness among people in creative professions, was swallowed without question, and was consequently taken as unequivocal evidence for the right brain as the seat of creativity. Handedness, it turns out, is not a reliable metric for lateralization. For instance, although left hemisphere speech is found in 96 percent of right-handers, this is also true of 70 percent of left-handers.[40] Indeed, the neurophysiological foundations of handedness and language lateralization are not the same.[41] It has in fact been suggested that handedness has only a modest relationship with language and that the key factor in relation to cerebral asymmetries is actually *praxis*, which refers to controlled, skilled manual movements and gestures.[42]

Bogen's ideas on the creative right brain functions and the inhibitions of the same via the controlling left brain provided prime fodder for the many dubious yet ubiquitous assertions that we can readily identify in common discourse even today, like the left hemisphere-dominated western mind versus the right hemisphere-dominated eastern mind, the advocacy of the use of New Age practices (in the West) like yoga and meditation to release the true potential of the right hemisphere, and the railing against left hemisphere-focused pedagogical practices in education.[43] The surprising thing about these developments is that Bogen did little to hide the fact that his assertions were largely an exercise in speculation. This is apparent even in the grounds for his choice of term to characterize right brain functions (as an apt contrast to that of the "propositional" left brain): "We would do well therefore to choose arbitrarily a word, homologous in structure with the word 'propositional' but sufficiently ambiguous to permit provisional use. For example, we can say that the right hemisphere has a highly developed 'appositional' capacity. This term implies a capacity for apposing or comparing of perceptions, schemas, engrams, etc., but has in addition the virtue that it implies very little else. If it is correct that the right hemisphere excels in capacities as yet unknown to us, the full meaning of 'appositional' will emerge as these capacities are further studied and understood. The word "appositional" has the essential virtue of suggesting a capacity as important as "propositional," reflecting a belief in the importance of right hemisphere function."[44]

Fifteen years after these early formulations, the continued influence of these ideas in the academic world was evidenced at the 140th Annual Meeting of the American Psychiatric Association, which took place in Chicago in a symposium on "Hemispheric Specialization, Affect, and Creativity." Papers from this symposium were subsequently published in a special issue of the *Psychiatric Clinics of North America* journal.[45] The paper authored by Bogen essentially constituted a summary of his views as presented in 1969 and included little in the way of new evidence. A notable conceptual advance was offered in the paper, however, and this was the idea that both hemispheres differentially contributed to the four different stages of creative problem-solving: preparation (build knowledge to understand the problem), incubation (take time away from the problem), illumination (the sudden arrival of a novel solution), and verification (check for the adequacy of this new solution to the problem).[46] Right hemisphere operations were proposed to play at least a partial role during the incubation phase, whereas the preparation and verification stages were believed to primarily involve left hemisphere operations. The corpus callosum was thereby thrust again into the spotlight as the brain structure of central importance in the context of creativity: "What is required is a partial (and transiently reversible) hemispheric independence during which lateralized cognition can occur and is responsible for the dissociation of preparation from incubation. A momentary suspension of this partial independence could account for the illumination that precedes subsequent deliberate verification."[47]

So, by the late-1980s, even the most prominent advocate of the power of the right brain put forward a deliberately balanced view that emphasized the dynamic interaction between both hemispheres as central for optimal creative ideation.[48]

But to little effect. The creative right brain myth/truth had already settled into the collective consciousness and proved impervious to revision.

Empirical Evidence for the Creative Right Brain

Much of the early experimental evidence for the unique role played by the right hemisphere in creativity was indirect at best, and followed approaches that were based on the inference that the direction of head or eye movements that took place when a person was engaged in a specific type of thought reflected the hemispheric lateralization associated with that activity.[49] So, gazing toward the left while engaging in creativity-relevant operations would be suggestive of the involvement of the right hemisphere in creative thinking. Studies using paradigms that followed this rationale began to be published in the 1970s and several suggested a special role for the right hemisphere in creativity.[50] However, other scholars have since questioned the accuracy in the logic applied to reach such conclusions in some of those studies.[51]

The application of electroencephalography (EEG) in the study of creativity allowed for a far more direct test of the creative right brain hypothesis, as the technique involves the recording of electrical brain activity using scalp electrodes. Colin Martindale (1943–2008) was the first scholar to champion EEG approaches in creativity. His early studies left little doubt as to what his stance was in relation to the creative right brain idea, given that he only recorded EEG over right hemisphere regions.[52] His body of work was nonetheless vital in the study of the creative brain. For instance, he was the first to pinpoint how the pattern of specific brain rhythms that were associated with creative thinking varied as a function of the stage of creative problem-solving. He also examined individual differences in creativity. Following on from a series of three EEG studies, where brain activity was recorded in both hemispheres, one of the prominent conclusions he put forth in an influential paper was that right hemisphere activity is greater in highly creative participants during creative thinking. Upon close reading, however, it becomes clear that this statement in the abstract was an oversimplified account of the actual findings detailed in the paper, which were far from straightforward.[53] Nonetheless, findings in this vein were advocated as lending support to the creative right brain hypothesis well into the late 1990s,[54] although the claims were challenged from the get-go.[55]

The alternative view is the hemispheric balance or integrated hemispheric function idea. Here the "confluence" of the actions of both hemispheres is held to be vital for creativity.[56] Strangely enough, both the creative right brain and hemispheric balance camps can draw on much of the same evidence to assert their differential views. The key difference lies in what aspect of the evidence is emphasized. For instance, on the one hand, the view that the right and left hemispheres subserve primary process cognition (free-associative, analogical, concrete) and secondary process cognition (reality-oriented, logical, abstract) respectively, and that the former is purportedly more important for creativity and is accessed more effortlessly by highly creative individuals, is typically drawn on to make a case for the creative right brain hypothesis.[57] On the other hand, the fact that these two modes of thought—as originally formulated by Ernst Kris (1954)[58]—were used to explain individual differences in creativity, such that highly creative people are better able to alternate between the two modes, can also be taken as the theoretical rationale underlying the integrated hemispheric function idea.[59] Even Martindale suggested that "the difference in opinion may be a semantic one."[60]

In the decades that followed, the creative right brain hypothesis was extensively explored indirectly by means of behavioral studies,[61] directly through brain-based studies,[62] and via a combination of both.[63] The findings, however, are often difficult to compare and integrate, as the metrics vary widely from study to study. These include the type of right hemisphere functional engagement vis-à-vis the left hemisphere (relative or absolute dominance), the level of creativity of the sample (based on individual differences: high/low; or group membership: artists/nonartists), and the type of brain measure (event-related activation, resting state, anatomical connectivity, and so on). The situation is such that, whichever camp you belong to, you can find evidence for your preferred view. There are studies that show right brain dominance,[64] left brain dominance,[65] and bilateral hemispheric engagement[66] in relation to creative activities. Evidence that speaks in favor of right brain dominance for visual creativity[67] is matched by findings against the same view.[68] Little wonder then that even meta-analyses and reviews on the topic put forth opposing conclusions,[69] as they sample only partially overlapping literatures and sometimes even misconstrue the findings of the original studies.[70] The only reliable conclusion that can be derived, given all that we know so far, is that any crude notion of the right brain being more important for creativity is simply untenable. Creativity necessarily "entails cooperation among many different cerebral areas and involves both hemispheres."[71]

Contemporary Views on Left versus Right Brain Function

The problem with the continued focus on the creative right brain idea in the original form is not only that it is dated, but that it also draws attention away from more nuanced contemporary conceptualizations in relation to hemispheric asymmetries. For instance, instead of approaching the question at the level of uncreative versus creative thinking or verbal versus visual faculties, focusing on specific cognitive operations in relation to creativity has led to some interesting postulations that are circumscribed in scope and, therefore, testable. The proposal that the moment of creative insight—the sudden conscious dawning of a new idea experienced often as an "aha" or "eureka" moment—was lateralized to the right hemisphere gained popularity in the 1990s,[72] and was borne out by neuroscientific evidence in the early 2000s.[73] In contrast, other creative cognitive operations, such as conceptual expansion—or the propensity to widen knowledge structures to assimilate novel elements and thereby create new ideas— show the opposite pattern of lateralization to the left hemisphere.[74] Other scholars go beyond the cognitive domain. For instance, Alice Flaherty's work draws attention to a basic shortcoming in the literature, namely that right brain models place emphasis on the "skill to create" while disregarding the "drive to create."[75]

What are the strategies we can adopt to develop a more accurate understanding of hemispheric specialization in the context of creativity? There are at least three potential approaches. One way to clear the conceptual confusion is to distinguish between assertions of experimentally derived asymmetries (e.g., linguistic left hemisphere versus nonlinguistic right hemisphere) from those of philosophically motivated asymmetries (e.g., symbolic left hemisphere versus imaginative right hemisphere).[76] The former, after all, should be more reliable as they are based on empirical evidence, whereas the latter are largely speculative. However, the problem of dichotomizing functions persists because asserting two opposing frames across most contexts constitutes averring crude generalizations that stand to snuff out nuance, depth, and truth.

A second strategy is to widen one's focus outside the context of creativity to the hemispheric asymmetry literature on other psychological functions. Studies from recovering stroke patients, for instance, are useful in elucidating the constraints of hemispheric specializations,[77] as they suggest that asymmetries manifest as a function of degree rather than a binary.[78] In this view, dynamic compensatory mechanisms between the left and right hemispheres facilitate most higher-order functions, particularly those involving problem-solving, language, and imagination. So, for instance, the neurotypical brain shows left hemisphere engagement in word retrieval, but right hemisphere engagement in word retrieval is evidenced following stroke.[79] Such evidence is suggestive of similar latent functionality present in both hemispheres that is drawn upon when needed. These findings have important implications, as they beg the question of how much we can really conclude about the veracity of left versus right hemisphere specializations and the dynamics of their interactions, especially given that we largely limit ourselves to examining a thin slice of creative ideation at any given point. Under these circumstances, the evidence we gain will not allow us to determine which hypothesis better fits the data.

The third alternative is to collate and contrast different ideas of hemispheric specialization to evaluate whether the creative brain idea is better subsumed under some alternative functional characterization. For instance, successful reasoning is held to occur through left hemisphere processes of generating explanations and inferences, and right hemisphere processes of monitoring and inhibition of information that is no longer valid.[80] This fits with the notion of the left brain as the interpreter[81] in contrast to the right hemisphere, which focuses on maintaining and even augmenting indeterminacy.[82] Again, the take home message here is that "successful real-world functioning requires the participation of both hemispheres."[83]

More nuanced ideas on lateralization as applied to the context of creativity derive from such approaches, and the two most influential contemporary scholars in this regard are Mark Beeman[84] and Elkhonon Goldberg.[85] Drawing from literature on language comprehension, Beeman proposed that the left hemisphere exhibits fine semantic coding in that it is responsive to dominant and context-relevant information, giving rise to the processing of literal meaning. The right hemisphere, in contrast, is characterized by "coarse" semantic coding, and thereby activates a wider semantic net that draws in secondary and distant meanings that result in some degree of ambiguity.[86] The right hemisphere therefore allows for the formation of more unusual associative connections, which are in turn likely to have facilitatory effects on creative ideation. Beeman's view, then, is in line with the creative right brain hypothesis.[87] Goldberg's focus is larger in scale, as it is not solely derived from research in the domains of language and semantic cognition. In his view, the role of the left hemisphere, on the one hand, is to facilitate cognitive routines. The right hemisphere, on the other hand, is sensitive to cognitive novelty.[88] Although he elaborates on the importance of the right hemisphere in creative ideation, he also emphasizes the necessity of left hemisphere contributions for the successful realization and implementation of an original idea. In outlining the importance of both hemispheres in creativity, he puts forward what is essentially an integrated hemispheric function model for creativity.[89]

Concluding Thoughts

The creative right brain idea began as a rebellion of sorts against the high esteem bestowed on left hemisphere functions. In countering the blunt dismissal of the right hemisphere, the importance of its functions in relation to the human imagination were avidly advocated. The original message, that the left and right hemispheres walked hand-in-hand in the context of creativity, was lost in the process. The initial emphasis on giving the right brain the recognition it deserved as an equal to the left brain, as opposed to its feeble twin, metamorphosed over time to the cultivation of the superior right brain idea. We started at "the right brain plays a vital role in creativity together with the left brain" idea and ended at "the right brain is the seat of creativity" idea; a place we still find ourselves in today.

It is a strikingly persistent notion: the idea that the verbal-logical-narrative mind is distinct from the musical-artistic-creative mind, and that the distinction is maintained at the level of brain structure and function in a simple and tidy fashion.[90] It doesn't seem to matter that the evidence on hand suggests a stupendously complex picture. Exasperated "It's not that simple!" assertions by charismatic and ordinarily convincing experts are all but ignored. It is interesting to consider why this is. The natural inclination for the human mind to consider the world in terms of dualities, combined with the discovery of factual functional dualisms in relation to the most obvious and plainly visible division of the brain, had the effect of a permanent magnet. The division of anatomy and function along the longitudinal fissure became imbued with significance and purpose in a manner that went well beyond data or reason.

In some ways, this can be seen as a positive development. The advocacy for right brain functions was central to over-hauling the educational curriculum in the United States, as it emphasized the utility in holistic pedagogy by targeting so-called right brain skills via the use of more stimulating teach-ing styles and visuospatial techniques, all the while reducing the emphasis on rote learning. The focus on updating educa-tional practices in view of the importance of right brain func-tions helped accommodate individual differences in learning and paved the way in countering some of the systemic insuf-ficiencies that disproportionately affected culturally disadvan-taged and neurodiverse groups. In the words of Michael E. Staub: "Important insights can sometimes be based on imper-fect science."[91]

Jerome Bruner's stance on the powerful metaphor of the left hand may have been prescient in this context.[92] In his view, the left hand represented the faculties of intuition, spon-taneity, and imagination. The left hand (right brain) thereby symbolized "the dreamer," in contrast to the right hand (left brain), which symbolized "the doer." As interactions between antic left hand and practical right hand are necessary for cre-ative acts, imposing divisions between the two is to take a stance that is essentially counterproductive. In making a case (metaphorically speaking) for the integrated working of both hands, yet highlighting the particular power of the left hand, Bruner inadvertently reveals the grounds for the persistence of the creative right brain notion.

The catchiness of the idea of the creative right brain is perhaps best explained by the power it wields as a metaphor, one that captures the gist of our subjective experience and explanations about the world at large. If we perceive our world as unbalanced in being oriented toward edifying metaphorically left-brain logical functions at the expense of metaphorically right-brain imaginative functions, the clarion calls to rectify this imbalance depend in part on using this readily understood metaphor to efficiently make the case, spread the word, and effect change. The problems of misdirection and the perpetuation of falsehoods begin when the metaphor of the creative right brain is taken to be literal. And these problems are exacerbated when the right brain is touted as more important than the left brain for creativity. The seat of creativity is not limited to any one hemisphere. Creativity, by its very nature, demands an expansive throne.

2 Madness and Creativity

[H]is illness gave him that exquisite and painful poetic sensitivity, like living with one layer of skin missing. The only time he savoured was when the erratic pendulum of his moods swung through a period of hypomania. [Robert] Lowell called it his magical orange grove in the mindset of a nightmare. There his mind raced. He became fluid, a literary giant. He talked like a machine gun with blazing eyes.
—Excerpt from a 1998 BBC documentary film *Playing with Madness*[1]

It would be vain to try to put into words that immeasurable sense of bliss which comes over me directly [when] a new idea awakens in me and begins to assume a definite form. I forget everything and behave like a madman.
—Pyotr Ilyich Tchaikovsky (1840–1893)[2]

[I]f we trust Aristotle, no great genius has ever been without a touch of insanity. The mind cannot use lofty language, above that of the common herd, unless it be excited. When it has spurned aside the commonplace environments of custom, and rises sublime, instinct with sacred fire, then alone can it chant a song too grand for mortal lips: as long as it continues to dwell within itself it cannot rise to any pitch of splendour: it must break away from the beaten track, and lash itself to frenzy, till it gnaws the curb and

rushes away bearing up its rider to heights whither it would fear to
climb when alone.

—Lucius Annaeus Seneca (4 BC–65 AD)[3]

Unlike many of the myths/truths explored in this book, where
the seed of the notion can be traced back as germinating from
a particular context or the views of an identifiable few, there
is no shortage of people—including exceptionally eminent
scholars and creative artists—who have insisted on some form
of association between the states of mind that typify madness
and those of creativity. It is an ancient idea that has stayed
with us through the millennia. And, as with many beliefs that
continue to be maintained across generations, cultures, and
histories, while the particulars of the creativity and mental illness link have changed considerably over time, the basic shape
of the *Ur*-idea remains preserved.

The origins of this notion can be traced back to the archaic
concept of the "daimonic." Across several ancient religious and
mythological traditions, "daemons" were regarded as powerful
beings that possessed spiritual powers, capable of influencing
the thoughts, emotions, and behaviors of ordinary mortals.
They were often fluidly allied to traditional deities and were
simultaneously regarded as good (constructive) and evil
(destructive) forces.[4] So, on the one hand, the malignant focus
of such spirits was held to result in the unleashing of evil in the
world while also causing madness, lunacy, and insanity in
humans through daemonic possession.[5] Yet, on the other hand,
the benevolent actions of daemons could bring about divine
inspiration, profound insights, and creative flourishing.[6]

A reformulation of this idea was brought to prominence in
the works of Carl Jung and Rollo May, and later fleshed out

in detail by Stephen Diamond,[7] where a distinction is made between the "demonic," which is solely negative and evil, and the "daemonic," in which the destructive (fear, anger, rage) and the constructive (creativity, generativity, psychological growth) occupy two sides of the same coin. They are conceptualized as fundamentally coupled forces, with human potential as their common source. The evocative words of Rainer Maria Rilke capture this relationship perfectly: "If my demons leave me, I am afraid my angels may take flight as well."[8]

Indeed, this sentiment reflects what is frequently brought to mind when contemplating the notion of the "tortured genius"; a notion that, interestingly enough, tends to go in and out of fashion with the times. It is one that has been applied to describe British poets in the age of Romanticism (eighteenth–nineteenth centuries) and American writers in the modern era (latter half of the twentieth century). However, the Italian Renaissance (fourteenth–sixteenth centuries), a period of outstanding creative productivity, is not associated with a greater incidence of mental illness among artists.[9] Quite the contrary. That the vast majority of eminent artists of this period "were neither notably unconventional nor notably temperamental, but on the contrary studious, hard-working, courteous, sociable, and sophisticated,"[10] is evidence that that psychological disturbance cannot be a necessary condition of remarkable creativity. What is more, it compels us to consider the impact of environmental factors in the equation. In contexts of optimal social systems that nurture, support, promote, or encourage creativity,[11] is an association between madness and creativity to be expected?

In exploring the myth/truth of the madness-creativity association in this chapter, important definitional issues need to be confronted. What does it mean to be mad? How does this tie in with what it means to be creative? If the nature of this association has changed so much with time, are we all still talking about the same thing?

What Is Madness?

To address this question, we need to first identify what the word "mad" connotes. Our everyday use of the term indicates how expansive the concept is. Let us review some examples:

There was a *mad* rush at the airport.

It's *madness* to declare that schools be closed at the first sign of snow.

You'd have to be *mad* to spend $75 on a burger.

I'm *mad* about her.

I'm *mad* at her.

She is *mad*.

A *mad* man is walking naked down the street.

The last two examples stand apart from the rest, as they do not merely pertain to the perception of madness, but also the state of being mad. What's more, there is some degree of permanence assigned to the atypicality and dysfunction. It is therefore not a temporary or fluctuating state (as can be induced under the influence of psychoactive substances, for instance), but a trait or disposition signifying a larger pattern of an enduring condition.

Incomprehensible. Unreasonable. Irregular. Abnormal. Over-the-top. Impassioned. Agitated. Frenzied. These are the dominant characteristics associated with the condition of madness, and they suggest that two essential features combine to capture the meaning of the word: a cognitive component of behaviors or situations that are at some level perplexing (or at least beyond the understanding of the onlooker or viewer), accompanied by a conative or emotional component of a roused or excitable state.

Our present-day view of madness has come a long way.[12] Its earliest associations, as the manifestation of being possessed by evil spirits, can be traced back to at least the Neolithic period. The practice of trepanning, which involves the drilling or scraping of a hole into the human skull, was employed as a means to let out the evil spirits in an effort to treat mental disorders.[13] Madness was later believed to be the most dire form of punishment that was meted out by the Gods to mortals in early religious myths and heroic tales, either as a consequence of egregious conduct or simply due to the cruel hand of destiny. A later distinction was made between "good madness" as embodied by prophets, saints, mystics, and visionaries, which was poles apart from diabolical madness as orchestrated by Satan and propagated by heretics and witches.

The established practices of Hippocratic and Galenic medicine that were congruent with fifth and fourth centuries BC Greek philosophy drew on the balance of bodily fluids (the four "humors": blood, phlegm, choler or yellow bile, black bile) to explain health and illness. Melancholy was associated with an excess of black bile, for example. This constituted the earliest instantiation of the formal medicalization of madness. Over the centuries that followed, the practices of how to recognize and treat mental illness ranged from callous disregard (by means of formal segregation and abysmally wretched forms of institutionalization) to brutal interventions (through extreme neurosurgical practices such as lobotomies). Interestingly though, even under asylum conditions, and long before the establishment of the art therapy as a formal practice, patients were often allowed and even encouraged to express themselves by engaging in visual artistic pursuits. The artworks were scrutinized to determine how the produced features tied with the pathological conduct on display, as well as to gain some insight into the unknown workings of the patients' minds.[14]

Seventeenth-century philosophy saw the active association of madness with a lack of reason or irrationality; a fault in cognition as opposed to being rooted in demons and humors. In the words of John Locke (1690): "Mad men put wrong ideas together, and so make wrong propositions, but argue and reason right from them."[15] The parallels between the overactive imaginations of the mad and the poetic fueled the romanticization of that association in the Elizabethan era, so much so that it was a compliment to call a poet mad.[16] The famous words of John Dryden (1681) spun this association in yet another direction: "Great wits are sure to madness near allied, / And thin partitions do their bounds divide."[17] This heralded a novel linkage of madness with what was long thought to be its very opposite—mental sharpness, intellect, inventiveness, and reason. The potential for the positive in relation to mental illness was echoed in Ronald Laing's assertion: "Madness need not be all breakdown. It may also be break-through. It is potential liberation and renewal as well as enslavement and existential death."[18] Among the many advocates of the "mad genius" idea,[19] Cesare Lombroso (1891) was perhaps the most avid, claiming common hereditary roots for madness and creative eminence,[20] which interestingly enough is partially supported by recent genetic evidence.[21] He outlined several dozen signs of this association including "hyperæsthesia": "If we seek, with the aid of autobiographies, the differences which separate a man of genius from an ordinary man, we find that they consist in very great part in an exquisite, and sometimes perverted, sensibility . . . it is highest in great minds, and is the source of their misfortunes as well as of their triumphs. They feel and notice more things, and with greater vivacity and tenacity than other men; their recollections are richer and their mental combinations more fruitful. Little things, accidents that ordinary people do not see or notice, are observed by them, brought together in a thousand ways, which we call creations, and which are only binary and quaternary combinations of sensations."[22]

The twentieth century was characterized by marked vacillations in our conceptualization of mental illness: from the denial of its very existence[23] to the reaffirmation of the reality of madness;[24] from the acknowledgment of the stark insufficiencies of the psychiatric establishment's ability to diagnose or treat conditions with any degree of consistency or efficacy through most of recorded history,[25] to the avowed dependence on the biomedical model in the present day for the same. Mental illness is now diagnosed the world over following the guidelines of either the *Diagnostic and Statistical Manual of Mental Disorders* (DSM) or the *International Classification of Diseases* (ICD).[26]

The terms "madness" and "mental illness" are widely used interchangeably (though the former is often deemed more stigmatizing or derogatory).[27] It is, however, important to distinguish them when coming to terms with the madness-creativity myth/truth. The fifth edition of the DSM (DSM-V) defines a mental or psychiatric disorder as "a behavioral or psychological syndrome or pattern that occurs in an individual, the consequences of which are clinically significant distress (e.g., a painful symptom) or disability (i.e., impairment in one or more important areas of functioning)."[28] Following this very general definition, the DSM-V outlines the identification criteria for 157 disorders, the vast majority of which cannot be subsumed under the term "madness"—certainly not in the sense it has been used since antiquity.

The conditions that typify the specific state of madness are marked by *psychosis*, where the symptoms are characterized in some form by a clear loss of contact with reality, such as is the case with hallucinations and delusions. So, the disorders that are typically associated with this state are bipolar disorder and schizophrenia,[29] and scholarship in relation to these conditions forms the foundation of the madness-creativity link. Schizophrenia is characterized by a fundamental and severe loss of touch with reality, where the symptoms can be categorized as falling into three categories: psychotic (hallucinations, delusions, thought disorder), negative (avolition, anhedonia, flat affect, impoverished speech), and cognitive (attention, concentration, memory problems).[30] In contrast, bipolar disorder is characterized by "unusual shifts in mood, energy, activity levels, concentration, and the ability to carry out day-to-day tasks . . . from periods of extremely 'up,' elated, irritable, or energized behavior (known as manic episodes) to very 'down,' sad, indifferent, or hopeless periods (known as depressive episodes)."[31]

The Evidence Surrounding the Madness-Creativity Link

The kind of empirical work that has most emphatically argued for a positive link between madness and creativity are "historiometric" investigations of people who have achieved creative eminence in any vital domain of human endeavor, including the arts, the sciences, public office, social activism, business, and so on.[32] This approach involves the systematical analysis of the available biographical information in relation to these people, who are often referred to as geniuses and are regarded as being uniquely talented, to evaluate whether a higher prevalence of psychopathological traits and mental illness is evident in such accounts. The general verdict following this approach is mostly unanimous (yes to the madness-creativity link), but there is less common ground in relation to the particulars (yes to the link, but only in relation to a specific disorder and/or a specific artist category). For instance, evidence for the specific association between bipolar disorder and creativity mainly comes from the examination of prevalence rates of mood disorders among creative writers.[33]

In the words of Arnold M. Ludwig who examined more than a thousand extraordinarily creative people across different professions: "My findings show consistently and clearly that members of the artistic professions or creative arts as a whole—architecture, design, art, composing, musical entertainment, theater, and all forms of writing—suffer from more types of mental difficulties and do so over longer periods of their lives than members of the other professions. However, no single pattern of mental illness characterizes all of the creative arts professions . . . members of those creative arts professions that rely more on precision, reason, and logic (e.g., architects, designers, journalists, essayists, literary critics) are less prone to mental disturbances, and those that rely more on emotive expression, personal experiences, and vivid imagery as sources of inspiration (e.g., poets, novelists, actors, and musical entertainers) are more prone."[34]

Although depression is the mental disorder that has been dominantly associated with heightened creativity since the time of Aristotle,[35] the emphasis has shifted in contemporary times to the spectrum of conditions that are subsumed within bipolar disorder and schizophrenia.[36] While both disorders are typically regarded as distinct clusters of mental illness,[37] it is important to note, particularly given their association to creativity, that they show considerable overlap in terms of symptoms, neurobiological markers, neuropsychological profile, risk genes, familial patterns, and so on.[38] Arguments that have been put forward to consider these disorders along a single common dimension are longstanding.[39] Current proposals argue that "schizophrenia, schizoaffective disorder and bipolar disorder are in the continuum of severity of impairment, with bipolar disorder closer to normality and schizophrenia at the most incapacitating end."[40]

Unlike historiometric investigations, examinations of the madness-creativity link that employ the converse approach (ascertaining the likelihood of people with severe mental disorders holding creative occupations, as recorded in population registries) have delivered quite specific findings.[41] An examination of data from well over a million patients revealed that out of the eleven examined conditions (including schizophrenia, schizoaffective disorder, and unipolar depression), only bipolar disorder was associated with an enhanced, albeit modest, representation in creative occupations.[42] Population registry studies have also confirmed creative profession-specific findings from the historiometric approach,[43] namely that writers were especially vulnerable to almost all the examined disorders, and schizophrenia and bipolar disorder in particular, as they were more than twice as likely than controls to be diagnosed with those conditions.

The third approach investigating creative performance using psychological tests of creativity in people diagnosed with schizophrenia or bipolar disorder have delivered mixed findings that are challenging to interpret. The murkiness is owed to wide variations in (1) tasks employed to measure creative thinking across a limited number of studies, (2) sample characteristics and standards adopted when recruiting clinical populations, and (3) ambiguity in the reporting of the findings.[44] This means that at the level of single studies one can find evidence for whatever one seeks to find: enhanced creative thinking,[45] impoverished creative thinking,[46] or no association at all[47] between creativity and these mental disorders. Recent meta-analyses of the extant literature have revealed that, on balance, people diagnosed with either bipolar disorder or schizophrenia do not show enhanced performance on measures of creative thinking such as divergent thinking and creative cognition tasks.[48] In fact, the opposite is true in the case of schizophrenia, as the disorder is associated with poorer creative performance overall.[49]

The one persistent theme that emerges across all three empirical approaches—historiometry, population registry analysis, and experimental psychology—in the study of the madness-creativity link is that a further consideration needs to be factored in: that of a continuum between psychosis and good mental health.[50] This is rooted in the observation that any consistent advantages in creative function reported in relation to such disorders do not pertain to the clinical level. They are instead limited to the "subclinical" level,[51] as gains are often found in individuals with a higher-than-average risk of developing these disorders. These include family members of people with psychoses, and neurotypical people with a higher-than-average degree of psychopathological traits. Population registry studies certainly attest to this idea; nondiagnosed first-degree relatives (parents and siblings) of people with bipolar disorder, schizoaffective disorder and schizophrenia are overrepresented in creative professions.[52] Meta-analyses of task-based findings also confirm enhanced creative thinking faculties in subclinical populations,[53] as people at risk for bipolar disorder and those with a high degree of certain schizotypal personality traits perform better on a wide range of creativity tasks.[54] Vulnerability to bipolar disorder and positive schizotypy has been characterized as embodying approach-based motivation that is essentially driven by novelty seeking,[55] which in itself is a personality trait that is associated with greater creativity.[56] What is more, not only do creative artists demonstrate higher degrees of such schizotypal traits,[57] their art making and aspects of the creative process that include the "flow state, altered experience, internal dialogue, vivid imagery, distractibility, introspection and high self-esteem" are influenced by positive schizotypy.[58] Shared phenomenological features between creative states and bipolar disorder include heightened moods leading to expansive thinking; activating high energy feelings co-occurring with racing, urgent, and determined thought flow; and reciprocal action between mood and creativity.[59]

To summarize then, in reviewing the vast and disparate repositories of evidence on this topic so far, we can glean the following clear patterns.

1. People with severe psychosis generally either perform no differently or worse than control groups on tasks that assess creativity-relevant psychological operations.

2. In contrast, better performance on such measures is demonstrated by first-degree relatives of people with severe psychosis, as well as neurotypical people who exhibit a high degree of select psychopathological traits but at a subclinical level (schizotypy).

3. Creative artists generally report a higher degree of schizo-typal traits.

4. A high proportion of creatively eminent people exhibit severe mental disturbances.

5. Of all the mental disorders, only people with bipolar disorder are slightly more likely to hold creative occupations.

6. Of all the creative occupations, only writers show an increased likelihood of being diagnosed with mental illnesses.

7. First-degree relatives of people with bipolar disorder, schizophrenia, and schizoaffective disorder are more likely to hold creative occupations.

A Touch of Madness ≠ Being Imbued with Madness

As we delve further into the implications of such findings, let us return to the start of the chapter and reexamine the selected quotations more closely. The excerpts touch on a particular state of madness that is associated with the creative spark: the fearlessness in striding toward an undiscovered possibility space filled with the answers one seeks, the fervor that accompanies those journeys to momentous insights, the absolute confidence in the fulfillment of one's destiny, and the feeling of being in control and in command, being all-knowing and all-perceiving, inhabiting and operating in a sacred space apart from the rest.

There are three factors to bear in mind here that are vital in the consideration of the madness-creativity myth/truth: specificity, temporality, and phenomenology. First, the states of madness associated with creativity are highly circumscribed in terms of their characteristics—they mainly pertain to features of perceiving, feeling, thinking, and behaving that fit with symptomatology typical of manic states (one phase in bipolar disorder) and schizotypal states (one category of symptoms in schizophrenia). So, they are highly specific, even in relation to the aspects of bipolar disorder and schizophrenia they may concern.

Second, the wise minds across human history that have advocated for the link between madness and creativity have generally been referring to states of madness ("phasic" in that they ebb and flow; momentary), and not traits of madness ("tonic" in that they persist; enduring). This is a critical distinction. Lumping together two very different conditions—the occasional departures from reality (states) and a habitual withdrawal from the world as we know it (traits)—is not only erroneous, but also utterly confounds the picture and cannot but give rise to a skewed consideration of the madness-creativity myth/truth.

Third, while there may be phenomenological parallels between states and traits of madness, it cannot be assumed that these necessarily generalize to outcomes in an equivalent manner. After all, psychosis is marked by persistent reality distortion, whereas a touch or tincture of madness is just that—a mere hint; being slightly askew; not quite on the same turf, but with a continued foothold on reality. People with differing state and trait conditions will therefore diverge in how they filter the world, and in the consequences that result from their inspired moments.

Even if it may be the case that active "states of madness" often accompany creative moments, the same cannot be said of active "traits of madness." Having certain forms of mental disorder means seeing the world differently, and this perception may sometimes be distorted in a manner that translates to a deeper understanding of the world around us or the human condition and can lead to greater creative generativity. But this appears to be more the exception than the rule. To put it glibly—not all who are mad are creative, and not all who are creative are mad. There is no case to be made here for an all-or-none phenomenon.

The findings on at-risk populations and schizotypy are insightful in relation to the "state versus trait" discussion as they point to the consideration of the middle ground, namely that states of madness occur more often in those with a "predisposition" to traits of madness. Some evidence on the phenomenology of the creative process,[60] as well as parallels in cognitive biases exhibited by artists and people with high levels of psychopathological traits, such as latent disinhibition and overinclusive thinking,[61] suggest that this is the case.[62]

The idea of shared vulnerability between creativity and psychosis[63] is longstanding, as are the attributed causes to evolutionary principles.[64] The gist of the notion is that, given the severe debilitative impact psychosis has on the lives of the afflicted individuals, such forms of mental illness would have been weeded out through the selection processes of evolution had the condition not conferred some advantage to our species. This advantage is believed to be that of enhanced creativity. The effect of such shared vulnerability is that it takes the form of an inverted-U function. Too much or too little cognitive control is not conducive to creativity, as they lead to rigidity or chaos respectively (the latter being the case in diagnosed psychosis). The middle "Goldilocks" ground of neither too much nor too little cognitive control, thereby possessing the optimal level of flexibility (as seen in at-risk groups and schizotypy), is held to be especially conducive to the creative ideation process.[65]

The Chicken or the Egg: Which Came First—The Illness or the Creativity?

The fact that there are other variables to consider in relation to the madness-creativity myth/truth is illustrated by the finding that creative writers are far more at risk of being diagnosed with not only bipolar disorder and schizophrenia, but also several other severe conditions such depression, anxiety disorders, alcohol abuse, and drug abuse. Writers even show an increased risk to commit suicide.[66] Findings like these that are limited to specific professions indicate that shared genetic or psychological vulnerability alone cannot explain the pattern of association between psychoses and creativity. We need to consider what other factors put some creative artists, such as writers, more at risk.[67]

One piece of the puzzle comes from a recent population register analysis of more than four million people, which showed an increased risk among students enrolled in degree programs related to the fine arts or applied arts of developing schizophrenia, bipolar disorder, and unipolar depression *later* in adulthood.[68] This suggests that there is more to consider about the context of undertaking creative professions that poses an enhanced risk of developing mental disturbances.[69]

Data from the US Census Bureau suggests that there is a great degree of risk that comes hand-in-hand with following creative career paths.[70] Artists are three times more likely to be self-employed than other workers. For instance, while the average self-employment rate of all workers between 2012 and 2016 was 9 percent, 34 percent of all artists in the United States were self-employed. Artists are also less likely to be in full-year or full-time employment, and this is believed to partially account for why their annual median incomes are lower than those of other workers with similar education levels. Artists are also more likely to work in the private sector. Data from Canada over the 2001–2009 period shows that part-time jobs in the private sector are associated with the highest degree of precarity (around 70 percent) relative to all other sectors (around 30 percent or below).[71] This was most recently illustrated during the COVID-19 pandemic, the repercussions of which were declared "a disaster for artists" by the RAND Corporation as unemployment rates for this professional category skyrocketed, particularly for performing artists.[72]

Another level at which precarity occurs is during creative ideation itself, as it is a process that is low in reliability (one cannot consistently prompt a creative idea on cue) and low in validity (the generated ideas are not necessarily creative at each attempt).[73] So, in developing creative ideas and products, an artist cannot count on her thought processes to deliver what she wants, when she wants it. Many an artist has lamented the enormous pressures that come with the expectation of producing a new creative work following the success of a previous project.[74] Risk is often involved in the generation of a creative work, and as creativity-relevant thought processes are challenging to employ, they can result in mental conflict and tension owing to the emotional and intellectual stress involved.[75] The added burden of the reception of the artwork also needs to be factored in, as the artist is at the mercy of the field and the public to be deemed a success. So, there is much about the context of the creative ideation process that is outside of the control of the artist.[76]

When exploring the case for the madness and creativity association, the focus has largely been on the tumultuous inner worlds of creators or the turbulent outer world against whose status quo artists react.[77] Far less ink has been devoted to the inherent instability of creative industries, combined with the inherent precariousness of the creative process in the vast literature on the subject. And this has led to an inadequate understanding of the nature of this association.

The striking words of Petteri Pietikainen convey why it is important to include the consideration of such factors in the equation: "Nobody is born insane. The mad, the mentally ill, the insane were, in most cases, not destined to lose their minds. If their lives had taken different paths, most of them might not have been burdened with mental illness."[78]

Concluding Thoughts

The feeling of being in flow, out of step with the world yet in step with one's own beat, inspired, immersed, energetic, frenetic, brimming with ideas, deeply trusting of one's vision no matter how much it departs from the known or the acceptable, unperturbable by the outside forces of the real world, of communicating with oneself with sparkling clarity, feeling the unshakeable confidence that one knows things that others cannot hope to understand: all of this and more are part of the complex phenomenological experience of ideational generativity. And oftentimes these states correspond very closely to those experienced in states of furore and mania.

In the section of *A Midsummer Night's Dream* that commences with the following lines—"Lovers and madmen have such seething brains, such shaping fantasies, that apprehend more than cool reason ever comprehends. The lunatic, the lover and the poet are of imagination all compact"—Shakespeare draws attention to where the similarities lie between these states. First, in the state of overexcitement or frenzy that accompanies their perceptions and fantasies—which are directed to different targets in the case of the madman, the lover, and the poet—and second, in their capacity to perceive the world in ways others do not.[79]

But therein ends the similarity.

Albert Rothenberg noted that "while the demarcation between psychotic thinking and creative thinking is superficially thin, the underlying psychological dynamisms are worlds apart," and that the thought processes involve entertaining the idea that "something is both true and not true at the same time is highly illogical and, on the surface, quite irrational. That two or more elements can occupy the same space is beyond the dictates of our experience, and it seems a formulation of the incredible, bizarre, and fantastic. Persons in the throes of psychosis do indeed have such conceptions as these, believe literally in them, and allow them to guide their behavior," whereas "the truly creative person is oriented toward producing something outside of himself, is rational, and is completely aware of logical distinctions."[80] This fits with the observation that when artists develop psychoses, more often than not one of two outcomes comes to pass—the quality of their work deteriorates or their productivity is stalled.[81]

The key insight to bear in mind here is that the eccentric mode—the ability to entertain incongruous thoughts simultaneously, generate unusual ideas, and so on—is only one side of the coin. The modes of overinclusive thinking, heightened sensation seeking, exceptional visuospatial imagery, expansive access to informational flow, and more,[82] are not sufficient conditions in and of themselves to attain creative productivity. A creative work, after all, is not only one that meets the definitional criterion of novelty or originality. It must also meet a second criterion of being a work "that is accepted as tenable or useful or satisfying by a group in some point in time."[83]

This is captured in Frank Barron's assertion that "the creative person is both more primitive and more cultivated, more destructive and more constructive, occasionally crazier and yet adamantly saner, than the average person."[84] The association between madness and creativity is only one part of a much more complex story. What matters is the balance of opposed dual forces. According to Barron, the foundational process in creativity occurs when "oddness of thought or feeling" is "coupled with an ability to reconsider and reformulate," and gives rise to "a socially communicable original meaning."[85] By only focusing on one side of the equation, we miss the big picture.

So why do we liken madness to creativity despite all the evidence that points to a decidedly more complex and nuanced picture? It is perhaps because of the simple reason that our attentional systems are built to involuntarily detect oddities in the world around us. The normal is not noteworthy and does not stand out, whereas the abnormal (psychosis) and the original (creativity) do. So, we automatically associate the two with one another following the principle of shared incongruity, and we infer erroneously that this surface similarity looms larger and deeper than everything else. We cannot but be charmed by the eccentricities of big thinkers whose ideas change the world.[86] It is a captivating pattern partly because it is a hopeful one. To be afflicted with madness is a burden like no other. The poetic justice of our imaginative minds demands an upshot—that greatness be granted for the price of suffering.

3 Psychedelic Drugs and Creativity

At first there was merely a vague play of light and shade, which suggested pictures, but never made them. Then the pictures became more definite, but too confused and crowded to be described, beyond saying that they were of the same character as the images of the kaleidoscope, symmetrical groupings of spiked objects. Then, in the course of the evening, they became distinct, but still indescribable—mostly a vast field of golden jewels, studded with red and green stones, ever changing. This moment was, perhaps, the most delightful of the experience, for at the same time the air around me seemed to be flushed with vague perfume—producing with the visions a delicious effect—and all discomfort had vanished, except a slight faintness and tremor of the hands. . . . The visions never resembled familiar objects; they were extremely definite, but yet always novel; they were constantly approaching, and yet constantly eluding, the semblance of known things. I would see thick glorious fields of jewels, solitary or clustered, sometimes brilliant and sparkling, sometimes with a dull rich glow. Then they would spring up into flower-like shapes beneath my gaze, and then seem to turn into gorgeous butterfly forms of endless folds of glistening, iridescent, fibrous wings of wonderful insects. . . . I was surprised, not only by the enormous profusion of the imagery presented to my gaze, but still more by its variety. . . . But in spite of this immense profusion, there was always a certain parsimony and aesthetic value in the colours presented. . . . Although the effects were novel, it frequently happened, as I have already mentioned, that they vaguely recalled known objects. . . . But always the visions grew and changed without any reference to the characteristics of those real objects of which they vaguely reminded me, and when I tried to influence their course it was with very little success. . . . The difference between the room as I saw it then and the appearance it usually presents to me was the difference one may often observe between the picture of a room and the actual room. The shadows I saw were the shadows which the artist puts in, but which are not visible in the actual scene under normal conditions of casual inspection. . . . I can, indeed, say that ever since this experience I have been more aesthetically sensitive than I was before to the more delicate phenomena of light and shade and colour.

—Havelock Ellis (1898), *Mescal: A New Artificial Paradise*[1]

It was ghastly. I don't want to repeat the experience. . . . I have never felt more ill in my entire life. . . . When asked whether I was experiencing any perceptual distortions, I said—I see nothing and I hear nothing. I just feel terribly ill.

—Friedensreich Hundertwasser on experimenting with psilocybin in 1959[2]

The question of what it feels like to undergo a hallucinogenic experience has been eagerly explored ever since published accounts of experimentations with psychoactive substances[3] began to emerge in the mid-1800s,[4] and it is one to which countless riveting responses have been documented over the ages following encounters with a range of substances, from age-old entheogenic drugs[5] to their modern-day artificially synthesized counterparts.[6] What is striking when perusing such observations is the manner in which the experiences vary between individuals. Reflections in the vein of Havelock Ellis receive far more attention and space in both the academic and popular literature. It is easy to see why. Such stories are more evocative and outlandish than their counterparts, which is why they tend to draw our attention and get selectively propagated (and thereby exaggerated) across all media avenues.[7]

This is then the first point of note, the nascent flag to pin on the myth/truth board of drugs and creativity, so to speak. Contrary to popular opinion, not everyone undergoes visionary experiences following the ingestion of psychoactive substances. It is interesting to explore why this interindividual variability exists. In the many studies that have examined the grounds for such individual differences, personality attributes like openness to experience, tendency to conformity, trait absorption, and suggestibility, among several others, are proposed to play a mediating role.[8]

There is of course also a great deal of interindividual variability even among people who report undergoing psychedelic experiences. However, some notable generalities seem to surface when examining the testimonies. When phenomenological consistencies occur, they tend to be at the level of perception, where the common elements include vividness and otherworldliness of the hallucinatory experience, an expansion of one's consciousness, a heightened awareness of extraneous details that are usually unnoticed, synesthesia or crossmodal sensory experiences, fluidity of space, distorted sense of time, and so on. Feelings of profound meaningfulness and personal significance, which also typically accompany such experiences, tend to be the facets of such states that linger long after the initial experience.[9] Indeed, the power of such experiences to fundamentally alter one's worldview is attested to by self-reports of the abandonment of atheistic views as a consequence of partaking in such episodes,[10] which tend to be suffused with elements that evoke decidedly spiritual and mystical associations. Little wonder then that such substances have been used across the world within indigenous traditions in service of collective spiritual and religious experiences as well as ritual healing practices.[11]

The lasting potency of such substances is reflected in the temporal duration of the subjective experience, which spans three phases: (1) the *psychedelic state* that directly follows ingestion of the drug (lasting for several minutes or hours) and may include peak or mystical experiences, (2) the *afterglow state* (lasting for days or weeks), and (3) the *long-term residual effects*.[12]

The key question to explore in the context of this book is whether such experiences give rise to advantages in the creative sphere. And if so, to delineate what form this hypothetical advantage takes in terms of the implicated temporal factors (during which phase do the positive effects occur?), dosage factors (how much exposure is necessary and sufficient?), and effect specificity (what type of advantages can be expected?). For instance, does creative ideation occur when immersed in the phase of the psychedelic state (immediate acute impact) or is it limited to the afterglow state (delayed acute impact) or the long-term residual effects phase (chronic impact)? How much drug exposure leads to alleged benefits—is a single exposure sufficient for lasting long-term impacts[13] or does repeated exposure engender additive gains?[14] And what aspects of our capacity to generate creative ideas get enhanced following such experiences?

One thing we know for certain is that the answers to many of these questions are as yet unclear. As we will come to see over the course of this chapter, much depends on the level of analysis that is adopted when posing this question, the working assumptions of what it means to be creative, and the motives that drive the investigations in the first place.

Phenomenological Parallels between Creative and Psychedelic States

The natural kinship between creative and psychedelic states is largely taken for granted in the literature. The specifics of how the two are related actually receives very little scrutiny as the onus by and large is directed at merely showing that one impacts the other. We can begin to explore this question by considering the nature of creative states, just as we have done for psychedelic states. What does it feel like to 'be creative'? This is a surprisingly overlooked question, but it is one that needs to be addressed when exploring the particular myth/ truth of the association between the ingestion of psychoactive drugs and their impact on creativity.

Regardless of the theoretical model of choice (usually the derivative of a continual iterative cycle between the stages of idea generation and idea exploration,[15] or a progression chain[16] through the stages of preparation, incubation, illumination, and evaluation), the stages involved in creative ideation do not neatly map onto the aforementioned phases of the psychedelic experience. This is because creative thinking involves a dynamic interplay between unconscious and conscious processes that are voluntarily and involuntarily wielded over a variable span of time. Critically, unlike in the case of drug-induced psychedelic experiences, purposive goal-directed action is a necessary component of creativity. James Kaufman makes the case succinctly: "We all have this dream of effortless creativity; of the muse speaking through us and making us immortal. But it is a myth. Creativity takes hard work, revision, failure, extensive knowledge and persistence."[17]

A more fruitful approach would perhaps be to reflect on which aspects of the creative process, if any, mimic the process of taking a psychoactive drug. A wide range of operations of creative cognition have been the subject of empirical study.[18] Of these, the experience of "flow" is the creative state that appears to have the closest parallels with the psychedelic state in terms of phenomenological properties. Compare the following descriptions.

Drawing on the work of Walter Pahnke in the 1960s and 1970s on LSD, Majić, Schmidt, and Gallinat outline nine elements of the *psychedelic peak experience* as including: "(1) a sense of unity; (2) the transcendence of time and space; (3) a deeply felt positive mood; (4) a sense of sacredness; (5) the noetic quality; (6) paradoxicality; (7) alleged ineffability; (8) transiency; and (9) persisting positive changes in different domains, including attitudes and behavior towards the self, others, life and the experience itself."[19]

When describing the *flow experience* Juliette Bowers states: "This peak experience involves a state of absorption where one experiences a balance between his expected challenge in the task and his ability level. There are currently nine identified components of flow: challenge-skill balance, clear goals, unambiguous feedback, action-awareness merging, concentration on the task at hand, sense of control, loss of self-consciousness, transformation of time, and the autotelic experience."[20]

Given the unmistakable commonalities between these descriptions, it is unsurprising that direct parallels are drawn between creativity and the psychedelic experience. However, there are grounds to be cautious when drawing generalizations from surface-level similarities. For one, the flow state is not interchangeable with creativity. Indeed, such experiences are also reported in situations that do not call for any form of creative response, such as during long-distance running.[21]

Moreover, states of absorption and temporal disengagement also accompany activities that involve relatively passive forms of imaginative engagement, such as gaming and fantasy role playing, using social media, watching films, and listening to music.[22] Many of the phenomenological features associated with creative flow and psychedelic experiences are therefore not exclusive to either of these states.

The Particulars of the Drug Taker

Much has been made of findings from empirical studies that demonstrate a much higher prevalence of hallucinogenic drug use among artistic populations[23] and the purposive use of such drugs by many artists to abet the occurrence of creative breakthroughs.[24] However, nationwide surveys of substance use across different professions indicate that while the incidence of illicit drug use and substance abuse disorder are certainly higher than average among professionals within the arts, entertainment, and recreation industry, workers in the accommodations and food services industry not only showcase much higher rates of drug use, but also occupy the top rung of the incidence table.[25] So, the mere fact of more common drug use in creative groups does not in and of itself constitute an explanation for their heightened creative performance.

It would perhaps be more fruitful to consider why it is that these drugs are taken in the first place. The use of such substances to enter a state of feeling inspired or to enhance one's cognitive abilities[26] may in fact be secondary to the primary aim of emotion regulation.[27] After all, such drugs give us the ability to alter or change our current mood in accordance with our needs in any given context.[28] Sometimes the aim may be to reduce anxiety and feel more emotionally stable. Other times it is to partake of a shared experience with someone else, engage in a spiritual moment, or just feel different. Among artists who use such drugs, self-reports often reveal a metacognitive or deep reflective awareness of when to use a specific psychoactive substance, and to what end.[29]

An even less acknowledged factor are the particulars of the drug taker, especially in the specific context of empirical studies that involve the ingestion of psychedelic substances. Contemporary scientific research on humans relies on the recruitment of volunteers who are apprised in advance about what the study entails. People are only allowed to participate in studies after they have provided their informed consent to do so. Safeguarding protocols exist for a good reason—to ensure risks are minimized and ethical conduct is maintained. What this means is that anyone who volunteers for a study on psychedelic drugs knows that they may be part of a group that undergoes drug use in some form over the course of the experiment. This raises the question then of how representative of the general population any study on psychedelic research can be, given that the openness to even consider taking such drugs is quite low in the general population. Recent estimates of prevalence rates of psychedelic drug use over the preceding twelve months among young adults hover around the 3 percent mark.[30] The problem of self-selection bias is therefore likely to be at play in extreme form in such studies,[31] and it is one that cannot be circumvented even when employing randomized double-blind experimental designs, the gold standard of the scientific method.

The issue of severely skewed sampling is a significant one that afflicts drug-related and clinical research in general. In investigations of the factors that differentiate such volunteers from the general population, personality traits have emerged as key distinguishing variables.[32] What is especially noteworthy in the context of this chapter is the significantly higher degree of openness to experience in such volunteers. This is relevant because openness is one of the strongest predictors of individual differences in creative thinking and is a key predictive factor in both artistic and scientific creativity.[33] In the few studies that have examined openness to experience and creativity in the context of drug use and behaviors, openness explained enhanced creativity over and beyond drug-relevant variables.[34]

So, there are a number of issues to bear in mind when attempting to derive meaningful conclusions from research in this field. First, consequential interindividual variability is associated with the psychedelic experience. Second, the sample under study in the case of research on the impact of psychedelic drugs on creativity cannot be said to be representative of the general population, as they are derived from a highly niche group.[35] This would be akin to generalizing findings about the physiological and behavioral effects of running to the population at large based on an examination of participants who clock less than eleven seconds on the 100-meter dash. Any conclusions that derive from samples with evident self-selection biases must therefore be tempered.[36] Third, given the unmistakable associations between creativity and openness to experience, levels of trait openness in such samples need to be assessed and factored into the data analysis of any study that investigates drugs and their impact on creativity. After all, if a person is naturally high in openness to experience and is predisposed to draw connections from seemingly unrelated information, it is tantamount to possessing an inherent readiness to generate original conceptual combinations. This leads to the following question. In what way is the effect of undergoing a positive and exalting psychedelic experience on creative thinking in such individuals comparable to those not characterized by such traits? A floodgate experience that takes one to unexpected realms may indeed nudge the former in the direction of serendipitous discovery over time, but it would be unwarranted to expect the very same consequences for the latter group.[37]

The Impact of Psychedelics on the Brain

Neuroimaging studies over the past few decades that have been undertaken to examine the question of the impact of psychedelics on the brain have not been conclusive. This is primarily due to the lack of similarity between the experimental designs used across studies, particularly in terms of drug type, dosage, and data analysis techniques. With that in mind, the one generalizable impact of ingesting psychedelics appears to be a reduction in both global cortical activation and functional connectivity. This effect is consistent across different drugs (e.g., LSD, psilocybin, ketamine) and the most commonly affected brain areas are the thalamus and the precuneus,[38] which is notable given the key role played by these regions in the regulation of operations relevant to consciousness.[39]

Hallucinogen-specific findings are also extremely revealing. For instance, a 2016 placebo-controlled study examining LSD-related brain changes showed a selective increase in cerebral blood flow to the visual cortex, a pattern that was positively correlated with the degree of complex visual imagery reported during the experience.[40] Decreased connectivity was found between the retrosplenial cortex and parahippocampus, a pattern that correlated with the subjectively reported experience of increased ego dissolution and altered meaning. Both these scene-selective brain regions are critically involved in spatial cognition functions, including navigation and memory.[41] Such findings attest to the fundamentally visual nature of the LSD experience in humans and the ripple effects from perturbations in that system (of one's location in the immediate visuospatial landscape) on wider self-based construals. The question of how such experiences play out in the visually impaired is attested to by a case study on a congenitally blind person who consumed LSD and other psychoactive substances over several years. Although the experienced hallucinations were never visual, parallel phenomenological properties were reported through the involvement of all other senses that together gave rise to deeply integrated synesthetic experiences.[42]

Such findings have informed theoretical models in the literature. One view envisions that the ingestion of psychedelics results in an odd coupling of unconstrained cognitive flexibility with an impoverished optimization of cause-effect information.[43] Another framework makes a strong case for an alternative type of induced imbalance in information processing during the psychedelic experience in the form of increased top-down or prior knowledge-driven processing, alongside decreased bottom-up or stimulus-feature-based processing.[44]

Neuroscientific studies on the impact of psychoactive drugs on animals are proving to be particularly fascinating. For instance, an increase in neural plasticity, both structurally and functionally, was shown to be consistently induced by different psychedelic compounds.[45] Psychedelics are classic serotonergic hallucinogens that act by galvanizing diverse cell populations including excitatory neurons, inhibitory neurons, and astrocytes across medial prefrontal and somatosensory cortices.[46] How such findings from animal experiments can be aligned with results from human brain-imaging studies is still unclear, particularly as the latter is strongly guided by the phenomenological aspects of the experience that are recorded through subjective self-report measures. But the fact that many of the questions regarding the neural consequences of type of drug ingestion (dosage amount and frequency of use on an unbiased sample) can be uniquely addressed in animal studies makes it a useful avenue from which to glean insights, which can be at least indirectly informative when considering the long-term and short-term perceptual, cognitive, and behavioral outcomes of the psychedelic experience in humans.

The Impact of Psychedelics on Creativity

Frank Barron, a pioneer in creativity research, was among the earliest scholars to investigate the influence of psychedelics on the minds of highly creative people whose impressions were recorded a day or two after their experimenting with psilocybin. He identified four commonly experienced elements: a perceived fluidity in the visual world, the novel appreciation of light and color, an increased sense of the soft beauty of the visual landscape, and absolute identification with an object when present in the visual field followed by utter unconcern when it was no longer so. Barron noted that while the psychedelic experience was profound for most people he examined, 10–15 percent found it unpleasant. Barron inferred from first-person accounts that what the experience seemed to do was to "dissolve many definitions and melt many boundaries, permit greater intensities or more extreme values of experience to occur in many dimensions" and he concluded that "certain aspects of the creative process, although by no means the creative process as a whole, are analogous to the kind of

breaking of perceptual constancies that is initiated mechani-
cally by ingestion of the drug."[47]

Fast forward to the present day. A systematic review of the
studies investigating the influence of psychoactive drugs on
creativity was published in 2017 and included both empirical
papers and case studies.[48] The key conclusion from this work
was that there is some manner of association between psychedelic drug use and creativity, but the nature of this relationship is far from clear. The evidence does not support a
direct link between the two (i.e., ingesting psychedelic drugs
does not enhance creativity per se), only an indirect one (i.e.,
ingesting psychedelic drugs enhances other factors that in
turn *could* positively influence creativity). Changes in artistic
style and aesthetic experience are quite consistently reported
by artists following the experience. The review also highlighted the necessity to consider how such drugs work differently across drug-naïve and chronic user groups (given, for
instance, higher substance use among artists).

Among the conclusions of concern were the numerous
methodological shortcomings in this field of research (some
of which have been highlighted in this chapter), which prevent a clear verdict from being delivered. Double-blind and
placebo-controlled studies are a true rarity in the examination
of the effect of psychedelic experiences on creative thinking.
This is a major problem, particularly when considering that
even in the absence of actual ingestion of such substances, the
mere elicitation of creativity-based expectancies in relation to
psychoactive drugs has been shown to facilitate creative ideation.[49] Indeed, the sheer expectation of ingesting hallucinogens in a context of a "psychedelic party" led to the reports of
drug-induced changes in 61 percent of participants in a pure
placebo study.[50] Expectations and context clearly matter in

this field of research and studies need to be designed to over-come these confounds.

An alternative strategy one can employ in order to come to terms with this area of research is to review the findings based on the phase of the psychedelic experience being examined (psychedelic state, afterglow state; long-term residual effects). While works of art created by artists while in the psychedelic state certainly show stark changes in style, such as the use of a more expressionist form and color,[51] they are not usually regarded as demonstrating more creativity. Instead, they are often rated as more imaginative,[52] even though the crafts-manship was clearly compromised. The question of whether these changes in style are attributable to changes in the way the world was perceived by the artist as opposed to merely the artifact of impoverished motoric dexterity induced under such conditions is an important one to consider.[53] Studies that have focused on alterations in specific cognitive operations have shown access to wider semantic associations when in the psychedelic state.[54] This fits with the findings that show a heightened propensity when in this state to attribute personal significance to stimuli that were deemed previously to be per-sonally meaningless or neutral.[55]

As far as long-term effects go, the limited evidence on hand from subjective reports of the participants suggests some long-term positive effects (afterglow state and long-term residual effects) in aesthetic sensibilities of artists as well as in relation to creative problem-solving more widely.[56] Stanley Krippner's surveys and interviews with hundreds of creative profession-als showed that although very few artists or musicians create works of art or perform when under the acute influence of the drug (i.e., the psychedelic state), most regarded the psyche-delic experience itself as having lasting artistic value.[57] This pattern fits with evidence that shows the positive effects on wellbeing and personal significance following a single dose of LSD in a safe setting persisted for one month (afterglow state) and twelve months (long-term residual effects) after the experience.[58]

From the empirical studies on hand, the seminal study led by Willis Harman in 1966 stands out, as it investigated the impact of creative problem-solving in a manner that took the stages of creative ideation and expertise into account.[59] The tested sample comprised professionals in the field of engineering, physics, math, architecture, furniture design, and commercial art. Their creativity and subjective experiences were tested before and after ingesting psychedelics. The participants had worked on problems that required a creative solution without success for many weeks prior to the experiment. The quantitative analyses showed an improvement in performance on the creativity measures from pretest to posttest[60] and the qualitative analyses indicated that 50 percent of the sample attributed their enhanced problem-solving in the illumination stage of creative ideation to the psychedelic experience. All in all, while there is some evidence from empirical studies of the positive influence of psychedelics on creativity, the size of these effects is modest at best and generally inconclusive given the methodological shortcomings and inconsistencies associated with these studies.

The importance of considering the phase of the psychedelic experience in relation to its influence on creativity is attested to by a recent double-blind and placebo-controlled study that examined the immediate and long-term effects of consuming psychedelics on creative ideation.[61] The ingestion of psilocybin led to a reduction in the number of ideas generated overall (lower ideational fluency) as well as in the number of unique ideas generated (lower ideational originality) when performing divergent thinking tasks. However, seven days later in a follow-up session where they were tested on the divergent thinking tasks again, the participants in the psilocybin group reported generating more novel ideas—ideas that were completely new to them—than the placebo group. While this subjective self-rating index is to be interpreted with caution, it is nonetheless interesting to note that this group experienced more moments of what can be called psychological, personal, or individual creativity (I-creativity),[62] a subjective experience of originality, in the afterglow state of the psychedelic experience, compared to a control group who were not in that state.

Concluding Thoughts

Aldous Huxley, one of the most famous advocates for the use of LSD made invaluable observations that are relevant to the myths/truths of psychedelic drugs and creativity in a 1960 interview in the *Paris Review*.[63] The following excerpts are particularly pertinent (66–68):

INTERVIEWERS

Do you see any relation between the creative process and the use of such drugs as lysergic acid?

HUXLEY

I don't think there is any generalization one can make on this. Experience has shown that there's an enormous variation in the way people respond to lysergic acid. Some people probably could get direct aesthetic inspiration for painting or poetry out of it. Others I don't think could. For most people it's an extremely significant experience, and I suppose in an indirect way it could help the creative process. But I don't think one can sit down and say, "I want to write a magnificent poem, and so I'm going to take lysergic acid." . . .

INTERVIEWERS

Would the drug give more help to the lyric poet than the novelist?

HUXLEY

Well, the poet would certainly get an extraordinary view of life which he wouldn't have had in any other way, and this might help him a great deal. But, you see (and this is the most significant thing about the experience), during the experience you're really not interested in doing anything practical—even writing lyric poetry . . .

INTERVIEWERS

. . . Is there any similar gain in psychological insight?

HUXLEY

Yes, I think there is. While one is under the drug one has penetrating insights into the people around one, and also into one's own life. Many people get tremendous recalls of buried material . . . the experience can be very liberating and widening in other ways. It shows that the world one habitually lives in is merely a creation of this conventional, closely conditioned being which one is, and there are quite other kinds of worlds outside . . .

INTERVIEWERS

Could such psychological insight be helpful to the fiction writer?

HUXLEY

I doubt it. After all, fiction is the fruit of sustained effort. . . .
To write fiction, one needs a whole series of inspirations about
people in an actual environment, and then a whole lot of hard
work on the basis of those inspirations.

In the final analysis, the experience of a psychedelic drug
on creativity is perhaps most accurately understood in the
wider context of its effect on other aspects of our psychological functioning. Of the people who have evocative and tran-
scendent experiences following the ingestion of such drugs,
some of them experience lasting changes in how they come
to appreciate new aspects of the world around them and, as
a consequence, themselves as well. This may have an indirect long-term effect on their creative output, just as on other
aspects of their lives. The bulk of the evidence suggests that
the immediate or short-term impact of such drug intake on
creativity is either negative or negligible.

The problem of skewed sampling is one that poses a sub-
stantially thorny issue for this field of study. This is not only
because the psychological makeup of people who tend to volunteer for such studies (i.e., being open and willing to experiment with such drugs) is not shared by a substantial proportion
of the general population. The fact that personality-based factors also vitally modulate the very nature of psychedelic experience must be acknowledged and taken into account. The
conclusions of any study are only as good as its design. And,
in this field of study, the experimental designs have this critical (and presently insurmountable) flaw in the recruitment of
participants that make up the sample under study.

The larger message from the enormous body of work in this
field is perhaps the most relevant of all—that such experiences

stand to introduce levels of deep and unexpected meaningfulness in the lives of many. This is because they allow for the
release of the ordinary constraints that are in place on our
conceptual knowledge. They open the mind to previously
unrecognized possibilities. A wider than usual access to information can serve many ends because it allows for the breaking of established patterns in the way we see the world, think
about the world, and act on the world.

The capacity of such experiences to spark awareness and
transform lives is very real. The therapeutic value of such
experiences is being increasingly recognized and strongly
advocated.[64] However, by overstating or understating the
case, we stand to lose out on harnessing the real power of
such experiences. The truth, after all, is necessarily nuanced.
The fact that the subjective effect of the psychedelic experience, as well as the level of psychological flexibility displayed
by a person, have been shown to mediate the therapeutic
response in alleviating symptoms of anxiety and depression
(and reducing tobacco consumption), should guide action
and advocacy in relation to such drugs.[65] Taking into account
the expectations and personality of the drug taker (set), the
atmosphere of the drug-taking session (setting), and the wider
environment from which the person comes from and returns
to (matrix) is necessary to make any reliable claims about
the potential of such drugs to deliver benefits to the human
experience.[66]

The guiding insight that should serve as a beacon when
navigating these trippy waters and selecting our approach
of examination and dissemination is as follows: the deriva-
tion of novel salience and personal meaningfulness, followed
by the incorporation of the same in some form within the
unique conceptual frames of our own minds, is the bedrock of

creativity and other peak psychological functions.[67] Agenda-based research stands to get in the way of consequential long-term progress in this sensitive field of study.[68] Getting to the heart of the matter may be best achieved by donning the spirit of Albert Hofmann, the first to synthesize LSD, whose reflections demonstrate an uncommon and invaluable blend of skills—openness, rigor, precision, and curiosity.[69]

4 Atypical Brains and Creativity

Defects, disorders, diseases, in this sense can play a paradoxical role, by bringing out latent powers, developments, evolutions, forms of life, that might never be seen, or even imaginable, in their absence. It is the paradox of disease, in this sense, its "creative" potential.

—Oliver Sacks (1933–2015)[1]

It would be a mistake to assume that the process of compensation always, without fail, ends in success, that it always leads from the defect to the formation of a new capability. As with every process of overcoming and struggle, compensation may also have two extreme outcomes—victory and failure—and between these two are all possible transitional points. The outcome depends on many things, but basically it depends on the relationship between (1) the severity of the defect and (2) the wealth of compensatory reserves. But whatever the anticipated outcome, *always and in all circumstances*, development, complicated by a defect, represents a creative (physical and psychological) process. It represents the creation and re-creation of a child's personality based on the restructuring of all the adaptive functions and on the formation of new processes—overarching, substituting, equalizing—generated by the handicap, and creating new, roundabout paths for development. . . . The positive uniqueness of the handicapped child is created not by the failure of one or another function observed in a normal child but by the new formations caused by this lapse.

—Lev Vygotsky (1896–1943)[2]

The disciplines of neurology and psychiatry tend to focus on atypicality in perception, thought, emotion, and behavior. Within the dominant medical model, atypicality is defined with respect to the agreed-upon norms of the time. Departures from the norm are defined in a manner that is essentially subtractive in that they are characterized as abnormal, defective, failing, limited, deficient, insufficient, lacking, impoverished, deviant, and so on. The blind spot of this form of quantitative approach, where functions are regarded as basically additive and linear, began to be recognized in the early 1900s as the approach allowed for neither a comprehensive nor an accurate picture of the nature of atypical function.[3] After all, deficiencies are not only marked by a loss of typical function but also feature the manifestation of novel ways of operating that come about as a function of adapting to atypical internal or external milieus. In the words of Vygotsky, "a child whose development is impeded by a defect is not simply a child who is less developed than his peers but a child who has developed differently." For example, "if a blind or deaf child achieves the same level of development as a normal child, then the child with a defect achieves this *in another way, by another course, by other means* . . . this uniqueness transforms the minus of handicap into the plus of compensation." Such insights called for an examination of atypicality that "has its own particular analytical objective" away from solely quantitative indicators.[4]

The need for alternative forms of scientific inquiry that pay heed to compensatory and generative processes[5] becomes especially obvious in situations where recognizable functional strengths and positive outcomes occur in the context of atypicality. For instance, profound loss in one sensory domain, such as when blindness occurs early in life, has been associated with enhanced function in other sensory domains when compared to sighted individuals. These include greater accuracy in sound localization,[6] superior tactile discrimination,[7] and better odor identification.[8] The advantages in the auditory domain also extend to absolute (or perfect) pitch: the ability to accurately identify a musical note in the absence of an external reference. This is a rare ability occurring in less than .01 percent of the general population,[9] so it is significant that congenitally blind children are four thousand times more likely than their fully sighted counterparts to possess absolute pitch.[10] Although it is more common among musicians (15 percent), especially those with early formal training before the age of nine,[11] absolute pitch is even more prevalent among blind musicians with early musical training (57 percent).[12]

This is an instantiation of the phenomenon of *paradoxical functional facilitation*, which is the puzzle of how "direct or indirect neural damage may result in the facilitation of behavioural functions."[13] Evidence for paradoxical functional facilitation has been documented in several aspects of psychological function, including child development, aging, learning and memory, expertise, self-based processing, epilepsy, recovery in neurological and psychiatric conditions, and creativity.[14] This facilitation can take one of two forms—restorative or enhanced. Restorative paradoxical functional facilitation occurs when normal or near normal function is attained following abnormal or subnormal levels of functionality, whereas enhanced paradoxical functional facilitation occurs when functionality in the affected person is superior to that of a neurotypical control individual.

Compensatory mechanisms via neural plasticity are held to be responsible for enhanced paradoxical functional facilitation. Neural plasticity or "neuroplasticity" is a general term that refers to the many different ways in which the brain changes "continuously throughout life, in response to everything we do and every experience we have."[15] Neural plasticity is enormously varied as it manifests at every level of the nervous system, from molecular to behavioral. Its impacts range from subtle to stark, and while its workings are typically geared to promote optimal functionality, neural plasticity can lead to suboptimal function (e.g., maladaptive learning in the context of addiction). In the case of paradoxical functional facilitation, plasticity is one of ten interrelated principles (or fundamental characteristics of the brain) that have been proposed to help explain how this phenomenon can come about at all.[16]

Staying with the example of blindness, the use of functional neuroimaging studies has provided evidence for plasticity at the level of brain activity patterns in the context of paradoxical functional facilitation in blind musicians. Sighted musicians with absolute pitch demonstrated the expected pattern of greater right-primary auditory cortex and cerebellar activity when engaging in a pitch discrimination task, whereas a blind musician with absolute pitch showed brain activity across both hemispheres in visual association areas as well as parietal and frontal regions.[17] Indeed, brain regions typically associated with visual processing undergo dynamic changes via neural plasticity as a consequence of blindness, such that these regions are co-opted for the processing of other sensory information as well as cognitive functions.[18]

Following the quotes at the start of this chapter, the phenomenon of paradoxical functional facilitation can be regarded in some sense as an instantiation of the innate creative impulse that is imbued within our human neurophysiology.[19] This chapter attempts to zone in on one aspect of this system:

creative expression, at the level of cognition and behavior, in the context of atypicality.

Born Atypical

No condition illustrates the phenomenon of paradoxical functional facilitation better than savant syndrome, defined as "a rare, but extraordinary, condition in which persons with serious mental disabilities, including autistic disorder, have some 'island of genius' which stands in marked, incongruous contrast to overall handicap."[20] Around 50 percent of people with savant syndrome have autism and the other half have other developmental disabilities. The incidence of savant syndrome in mentally disabled individuals is 1.4 per 1,000,[21] whereas anywhere between 1 in 10 and 1 in 200 people with autism are estimated to have savant skills.[22] The domains in which savant abilities are demonstrated include calendar calculation; language acquisition; mathematical skills; and memory, art, music, mechanical, or visuospatial skills.[23]

Distinctions are often made between three levels of savant skills. Most common are splinter skill savants who show an excessive preoccupation with and memory for music or sports trivia, historical facts, public transport schedules, maps, and so on. Talented savants, on the other hand, exhibit considerable musical, artistic, or other special abilities, such as in language or memory. Prodigious savants are rare and are characterized by the level of outstanding skills that meet the criteria of genius or prodigy among nonimpaired people. While the presence of savant skills in the presence of severe disability is a clear indicator that such skills are not limited to particular IQ levels, there is "a positive correlation between degree of exceptionality and more general intellective factors."[24]

Relevant variables to consider in this regard include enhanced perceptual functioning,[25] extensive and obsessive rehearsal,[26] motivational disposition to undertake repetitive actions,[27] and the ability to retain details and reproduce what is perceived.[28]

Among the most well-known prodigious savants in contemporary times is Kim Peek, who became famous when the world learned that he was the inspiration for the central character in the highly successful 1998 film *Rain Man*.[29] He was born with an enlarged head and showed developmental disabilities from an early age, displaying significant physical (sidelong gait), cognitive (problems with abstraction), and pragmatic difficulties (unable to button his clothes and carry out daily chores). Although he was unable to walk until the age of four, Kim began to memorize books by the age of eighteen months (as they were read aloud to him). He was able to read a page in eight to ten seconds and complete a book of average length in an hour. His photographic memory was singular; he could recall information verbatim from the texts he had read when prompted. At the time of his passing, Kim Peek is thought to have had a mental library of at least twelve thousand books and was considered to exhibit expertise to the level of a genius on fifteen subjects.[30] Scans of his brain revealed several abnormalities including a malformed cerebellum and a complete absence of the corpus collosum, the major tract connecting the left and right hemispheres, as well other white matter tracts such as the anterior and posterior commissures.[31]

It was long assumed that people with savant syndrome do not have the capacity for real creative thought and are limited to being essentially repetitive or imitative (albeit to an often astounding degree of virtuosity). But this is no longer considered a tenable view. Although savant syndrome is a condition that is defined at its core by the presence of fundamentally phenomenal memory skills, many savants progress beyond literal replication toward more creative forms of expression.[32] While Kim Peek's remarkable abilities were rooted in rote memorization, the transition to generativity followed in time. In the words of Darold Treffert (2011), "In Kim's case that evolution was from massive literal memorization, to comprehension and improvisation in the form of puns, to creativity in the form of insights and wit."[33]

One theoretical account has proposed that the basis of exceptional savant capacities arises from a breakdown in "concept-driven thinking," which is one of two perceptual modes that are in constant operation as we go about making sense of our environment.[34] Stimulus-driven or bottom-up processing derives from externally directed attention and allows us to pick up on the specific sensory features of the world around us. Concept-driven or top-down processing, in contrast, is fueled by internally directed attention and allows us to make sense of and understand the sensory information based on our knowledge and prior experience. A breakdown in the latter, which is held to occur in the case of savant syndrome, is proposed to give rise to an increased use of the former, which allows for atypical and augmented access to low-level raw sensory stimuli, the kind of information that under typical conditions would not be perceived at all. This "privileged access hypothesis" has been put forward to explain the information processing style that typifies savant syndrome (particularly where the developmental disability is autism), namely that the condition is marked by sensory hypersensitivity and an enhanced focus on detail, which can translate to an advantage in the unleashing of exceptional skills.[35]

Another account holds that elaborate "veridical mapping" best explains exceptional skills based on the pattern of evidence showing that the combination of savant skills accompanied by (a) superior perception (e.g., absolute pitch), (b) synesthesia (where sensory stimulation in one modality leads to an involuntary experience in another sensory or cognitive system), and (c) peaks of ability (e.g., hyperlexia: exceptional reading ability prior to comprehension) far more commonly occur in autism than in typically developing populations, suggesting isomorphisms at some fundamental level between these abilities. In essence, veridical mapping involves the functional repurposing of perceptual brain areas in service of higher-order cognitive processes, which enables "the enhanced detection and memorization of isomorphisms between perceptual and nonperceptual structures across multiple scales." It is this heightened retention "for couplings between homologous parts of similar patterns both within and across perceptual modalities that underpin the genesis and development of savant abilities."[36] There has been no direct examination of prodigious savants with and without manifest creative skills, so we cannot as yet ascertain what discernable differences, if any, are present in their neurocognitive profile that might account for the variance in exhibited creativity.

Suddenly Atypical

People who exhibit savant skills spontaneously from a very early age are classified as having congenital savant syndrome. There is a second category, however, acquired savant syndrome, which comprises individuals who develop savant skills later in life as a consequence of traumatic brain injury, despite having little to no prior interest in the domain in which they develop such capacities.[37] Well-known examples include Derek Amato, the musical savant,[38] and Jason Padgett, the mathematical savant.[39] Acquired savantism is extremely rare,[40] and the level of skills exhibited generally fall into the category of talented savants (as opposed to splinter skill or prodigious savants).

The novel emergence of specific artistic abilities, or de novo creativity, is also sometimes associated with the onset of a range of neurological disorders such as frontotemporal dementia (FTD), stroke, epilepsy, Parkinson's disease, and so on.[41] A tiny proportion of individuals with such conditions exhibit new artistic faculties following the compromise of neurological functions. Such findings have been most consistently observed in relation to FTD, a group of neurological disorders that is marked by the shrinking or atrophying of the frontal and temporal lobes. FTD symptoms are varied and may include personality changes, social behavior, and/or language difficulties, depending on the affected brain regions and the specific disorder type. Symptoms in behavioral-variant FTD include socially inappropriate behavior, apathy, and disinhibition, whereas gradual loss in language function occurs in primary progressive aphasia (PPA). It is in this latter group that the phenomenon of de novo creativity has been documented, mostly in the semantic variant PPA subtype (svPPA) where patients exhibit a continual loss of word meaning and comprehension but not word fluency, in addition to compulsive and rigid behavior. svPPA is associated with damage to the temporal lobes (particularly the anterior-most regions— the temporal poles—which are regarded to be the crossmodal semantic hub of the brain),[42] as well as limbic regions such as the insula, anterior hippocampus, and amygdala, with the relative sparing of frontal lobe regions.[43]

An unexpected and sudden interest and engagement in visual art was first noted in a small subset of patients with svPPA.[44] This was not a fleeting change as this novel behavior and interest endured over time. In fact, patients showed an intense preoccupation with their art and were compulsive in their manner of engagement.[45] The artistic style displayed in all cases was realistic and detail oriented, with an absence of symbolic or abstract elements. Although de novo skills have been mostly reported in the context of visual art, there is also evidence for them emerging in other domains like music[46] and writing.[47] While some scholars have questioned whether the emergence of such artistic skills in this manner can be classified as truly creative,[48] it is certainly true that the skill level demonstrated in such cases of de novo creativity do not approach the skill level seen in the case of savants.

Gradually Atypical

The facilitation of creativity-relevant faculties also occasionally materializes in slow and subtle ways over the course of the lifespan through the period of development in childhood and maturation into adulthood. These faculties have been noted in relation to neurodevelopmental disorders, chief among them dyslexia, autism spectrum disorder, and attention-deficit hyperactivity disorder (ADHD). In neurodevelopmental conditions, the disorder onset, which is marked by pronounced neurocognitive symptoms early on, occurs in childhood before puberty is reached and from there on follows a steady course.[49]

Despite their clear differences in key symptomatology, the fact that comorbidities often occur between neurodevelopmental disorders (for instance, the prevalence rates of comorbid ADHD with dyslexia and autism are 18 to 42 percent and 20 to 59 percent, respectively) suggests there are underlying psychophysiological commonalities in their vulnerabilities. Candidate factors for shared vulnerabilities that have been put forward to explain not only the primary deficit in such disorders, but also the high propensity for comorbidity between the disorders, include dysfunctions at the level of rhythm, timing, and synchrony in perception, as these factors are foundational to a range of psychological functions including attention, action, language, and social behavior.[50] For instance, dyslexia and autism show divergent patterns of bias in local-global processing on visuospatial tasks. Some studies indicate that dyslexics demonstrate enhanced global processing, whereas autism is associated with enhanced local processing.[51] So, is there convincing evidence of paradoxical functional facilitation in relation to creativity in these disorders?

Let us take the case of dyslexia first, which is a learning disorder characterized by a fundamental difficulty in being able to read and interpret what is stated in a text. It was certainly intuitive to expect that the compensatory mechanisms in dyslexia, which allow getting past the experienced language struggles, would result in an overreliance on visuospatial processing and, consequently, advantages in that domain.[52] Findings that indicate an increased incidence of dyslexic traits associated with practitioners of the arts seem to support this proposal,[53] but the empirical evidence for an information-processing bias in dyslexia that results in visuospatial cognitive advantages in general does not bear up. A review of the evidence on hand indicates that dyslexics on average actually perform worse than controls on visuospatial tasks.[54] Interestingly though, dyslexics also exhibited greater variability as a group—more extreme scores indicating higher and lower performance on visuospatial tasks—than their control groups, which would explain why individual studies differ so greatly in showing enhanced, impaired, or comparable function. With regard to performance on tests of creative potential, two independent meta-analyses have confirmed that dyslexia is not associated with higher creativity in children. However, the limited evidence on hand so far does appear to support some advantages in creative potential in the case of adults with dyslexia.[55]

There has been more empirical work on autism spectrum disorder, which is characterized by aberrant social behavior, language, and communication, alongside restricted and/or repetitive interests and activities. By some estimates, between a quarter and over a third of people with autism "show unusual skills or talents that are both above population norms and above their own overall level of cognitive functioning."[56] The question is whether these skills and talents are relevant to creativity. Although not everyone agrees,[57] evidence that suggests that they might be (at least in part) are promising. These include findings from the auditory domain showing that, for example, specific information-processing features that are more common in creative professionals, such as absolute pitch in musicians, are also more frequently found in nonsavant children with autism.[58] The pattern goes a full circle—people with absolute pitch show a higher degree of autistic traits[59] as do professional musicians with absolute pitch.[60] As for performance on tests of creative potential, a recent meta-analysis revealed that, when compared to neurotypical control groups, groups of people with autism exhibited greater originality on verbal divergent-thinking tasks.[61] In terms of creative achievement, while people diagnosed with autism are underrepresented in creative professions,[62] a comparison of visual artists and scientists showed that exceptionally creative visual artists exhibit higher degrees of autism-spectrum personality traits.[63] This latter finding ties in well with empirical work that demonstrates clear advantages on different aspects of visual perception in nonsavants with autism.[64]

Just as in the case of autism, people with a diagnosis of ADHD for which the primary symptomatology includes distractibility, impulsivity, and hyperactivity, are generally underrepresented within creative professions.[65] However, studies on nonclinical samples show that higher levels of ADHD traits are associated with greater divergent thinking as assessed by measures of creative potential.[66] The expected advantages in the case of ADHD are considered to be rooted in poor response inhibition, one of the signature cognitive deficits associated with this disorder.[67] Mild levels of disinhibition have in fact been shown to be advantageous in the context of creativity.[68] Indeed, there is some limited evidence to show children and adults with ADHD demonstrate advantages on select aspects of creative cognition that would benefit from distractibility. ADHD individuals were better able than neurotypical controls to ignore salient distractors that were designed to compel people to go down paths of least resistance and generate unoriginal lines of thought.[69]

What the limited research on the possibility of paradoxical creative functional facilitation in the context of neurodevelopmental disorders suggests is that yes—there is evidence for the same. However, the facilitative effects are highly specific in terms of the scope of their workings (mainly manifesting at subclinical levels of a particular disorder) and are highly circumscribed in terms of the degree of impact (mostly limited to modality-specific creative potential and not creative achievement as a whole).

Atypicality in Creative Artists

Several collected academic works explore the manner in which the trajectories of artistic productivity are affected by the onset of different types of neurological insufficiencies.[70] Examples include epilepsy in Gustav Flaubert and Alzheimer's disease in Carolus Horn, Federico Fellini's right hemisphere stroke and hemineglect in Lovis Corinth, Stendhal's aphasic transient ischemic attacks and migraine aura experiences in Blaise Pascal, Édouard Manet's ataxia and Willem de Kooning's dementia, and neuropathic pain in Frida Kahlo.

There are two striking generalities that can be gleaned from these necessarily subjective and speculative yet rich and informative examinations. The first generality is the extent to which artistic productivity continues *despite* brain disorder and neurological injury. Although reductions in the quantity and consistency of output are often observed, more often than not artists quite remarkably continue to make art following the acute phase of the experience of neurological deterioration. The cases where artistic productivity ceases are usually limited to when the severity of impairments is so extreme as to prevent basic functioning, such as in the case of severely debilitating pain (e.g., Ivan Turgenev), and when incapacitating conditions result in chronic institutionalization (e.g., Robert Schumann).

The second generality is the fact that changes in artistic expression and artistic style are often noted as well. Changes in artistic expression often occur when the preferred limb for creative expression can no longer be used, so adaptation to a new means of expression is necessary. For instance, Maurice Prost's amputation of his left arm and chronic neuroma following injury in World War I made it impossible for him to continue working as a goldsmith. Nonetheless he persevered in the arts by taking up sculpture and decided to devote himself to the enormous challenge (given his impediments) of carving animal sculptures out of hard stone. By having a self-designed pneumatic tool made that allowed him to overcome his handicap of being unable to use a hammer while his right hand held the chisel, he was able to regain his artistic autonomy and maintained his long-term creative generativity.[71]

Changes to artistic style are also known to occur following neurological insufficiencies.[72] However, the nature of the changes is not consistent from artist to artist. Some scholars have reported that the changes in style (including form, color, dimensionality, perspective) are such that the work becomes more abstract and less fractal,[73] whereas others note the exact opposite, with the work becoming more concrete and traditional.[74] A fascinating example comes from the longitudinal analysis of the artistic trajectory and neurological profile of the artist Anne Adams, which revealed a premorbid shift in expression from a generally concrete style to an increasingly abstract artistic mode. Then came a postmorbid shift to an artistic style that evidenced heightened photographic realism and vividness of detail, and this was followed by a reversal back to a generally concrete mode.[75] Changes in artistic style are particularly difficult to interpret because it is impossible to separate alterations that occur as a natural evolution of that artist's style from alterations that can be attributed solely to neural plasticity brought about by brain tissue degeneration or injury.

A deeper question, perhaps, is to ask: What accounts for this extraordinary resilience of artists in their capacity to create despite neurological damage? No doubt the role played by motivational factors such as the fundamental drive to create, as well as a sense of purpose, are an inevitable part of the answer. Another factor to consider is the built-in physiological resilience that comes from being highly trained for several years in their field of creative expertise. This hypothesis has been explored in the case of George Gershwin, who suffered a temporal right hemisphere cerebral tumor, and Vissarion Shebalin, who developed profound aphasia. Both artists exhibited an extraordinarily high degree of musical competence despite their severe neurological problems. In the case of Gershwin, his continued proficiency has been attributed to the preservation of musical representations in his left hemisphere, whereas Shebalin's resilience has been attributed to expanded cortical representations that come from extensive and intensive formal musical training that begins early in childhood.[76]

Concluding Thoughts

There is little doubt that the study of atypical brains in relation to creativity has brought to light the extraordinary power of the human mind in its capacity for resilience and continued generativity, through the emergence of compensatory processes and the blossoming of latent competencies. Nonetheless, the fact that positive outcomes are highly circumscribed, in that they are the exception rather than the rule, necessitates a discussion about the "when" and "why" of such manifestations. Under what conditions are such hidden talents likely to emerge and does the displayed generativity meet the requirements of what constitutes a creative act?

The central premise underlying the idea that the disrupted brain leads to enhanced ideational generativity builds on the possibility that "creativity may start by a disorganizing process, making links between remembered forms and structures which have not been previously linked together, and leading to 'newness.'"[77] If brain insufficiencies lead to the expression of dormant artistic skills, the question that is often posed is whether such capacities exist in everyone[78] or only in a minority of people.[79] That the latter is more likely to be true is attested to by the fact that paradoxical functional facilitation in the context of creativity is not only rare but also sporadic in occurrence. If only the former were the case, the phenomenon would be far more widespread. This is also attested to by studies of patients with Parkinson's disease where it has been observed that a small proportion of patients develop de novo artistic abilities following the intake of dopamine agonists for managing their condition.[80] The fact that this facilitation following medication is the exception and not the rule speaks against the idea that dormant artistic capacities are ubiquitous. It is also important to bear in mind that neurological insufficiencies are often associated with a decline in artistic output or deterioration of artistic style, as has been evidenced in relation to other neurodegenerative disorders like Alzheimer's disease.[81] The main inference that can be derived, then, especially in relation to the sudden appearance of novel artistic skills following neurological damage, is that it seldom occurs and therefore seems likely to transpire when the affected person already possesses latent capabilities that are conducive to artistic expression, and not otherwise. The continued generativity in artists (who possess manifest artistic skills) following the development of brain disorders also attests to this supposition.

Another point to consider is whether generativity following a compromise in brain functions is best characterized as "creative." Dahlia Zaidel makes an important distinction in this context between the drive to create versus the quality of the produced artwork by noting that "while the turning to art is itself innovative; the produced art, however, is not necessarily creative."[82] In a discussion on creativity in relation to

Tourette's syndrome, Oliver Sacks echoes this postulation: "I don't think one can have a purely creative disease. I mean in a way there is something brilliant about Tourette's in its stimulation of imagination and emotions. But control may be lost. And no work of art is possible without consciousness and control."[83]

A similar distinction was proposed by Morris I. Stein (1953) between the creative experience and the creative product in his now widely employed definition of a creative work as one that possesses the dual qualities of being both novel and satisfying.[84] The resonance of a creative work ranges in magnitude from being novel and satisfying to the experiencing self, where we adopt an *internal* frame of reference (individual level; I-creativity), to being novel and satisfying to the larger collective, where we employ an *external* frame of reference (historical level; H-creativity).[85] Brain insufficiencies may give rise to an increase in personally innovative behaviors or I-creativity that are rewarding and meaningful in those with latent generative capacities by heightening the drive to create.[86] H-creativity, however, additionally necessitates preserved executive capacities to generate, refine, persevere, and bring to life novel works that are designed to resonate as novel and satisfying, not only with oneself but also the larger cultural collective.

This chapter began with the words of Oliver Sacks, and it is to him (as the scholar whose compelling and evocative works demonstrated to the world at large the extraordinary significance of understanding atypical minds) that the final word is bestowed. In commenting on an article that linked Mozart with Tourette syndrome, he noted the following: "One must not romanticise Tourette's, or any other disease, nor make a romantic equation of disease and creativity (as Thomas Mann so often seemed to do). Creativity is usually in a different realm from disease. But with a disorder like Tourette's syndrome, especially in its phantasmagoric form, one may have the rather rare situation of a biological condition becoming creative or becoming an integral part of the identity and creativity of an individual. Whether this was the case with Mozart is unclear, but that it occurs among others, and often, is quite beyond question."[87]

5 Intelligence and Creativity

So long as a science knows only a fraction of the influences at work anywhere, it is almost as likely to lead astray as aright.

—Charles Spearman (1863–1945)[1]

Many believe that creative talent is to be accounted for in terms of high intelligence or IQ. This conception is not only inadequate but has been largely responsible for the lack of progress in the understanding of creative people.

—Joy Paul Guilford (1897–1987)[2]

I think I was made, as a child, to be far too self-conscious of my status as a "Termite" and of being smart and young for grade (all passive), and given far too little to actually *do* with this mental endowment (so I'd stop thinking about myself). My greatest regret is that my left-brain parents, spurred on by my Terman group experience, pretty completely by-passed any encouragement, of whatever *creative* talent I might have had. I now see the latter area as of *greatest* significance, and intelligence as its handmaiden. Sorry I didn't become aware of this fifty years ago.

—Sarah Ann Albright (study participant: Genetic Studies of Genius)[3]

To someone who asked Newton how he had managed to construct his theory, he could reply: "By thinking about it all the time."

—Albert Camus (1913–1960)[4]

There are few topics as singularly divisive as the study of human intelligence. Almost everything about it, from how to define and measure intelligence, to how to use its metrics to guide educational and social policy, is steeped in controversy. It may seem ironic that our ability to reason rationally and logically—the very hallmarks of abstract intelligence—about this topic cannot but lead to impassioned disputes where the wisdom, truth, humanity, and tact exercised on one side of the aisle are frequently called into question by the other side.[5] But the fact of the matter is that there is a fundamental divide in worldview at play in these discussions. While many regard abstract intelligence as the epitome of human abilities, others regard the examination of intelligence in the abstract as not particularly human. Indeed, the fact that artificial intelligence systems routinely prove far better than humans across a range of applications that draw on the kind of abstract intelligence abilities captured by IQ tests only serves to drive the point home even further. Both positions pursued in the extreme—overvaluing or undervaluing abstract intelligence—are associated with severe consequences: the eugenics movement, for instance, in the case of the former, and inadequate educational and societal provisions supplied for the most vulnerable people in the case of the latter.[6] There are no clear or convincing solutions on the horizon at present to deal satisfactorily with this impasse.

The challenge in the context of this chapter is to understand how intelligence came to be associated with creativity. In order to do so, let us first examine what the overlaps are between these two concepts.

A consensus of fifty-two experts in 1997 defined intelligence as "a very general mental capability that, among other things, involves the ability to reason, plan, solve problems, think abstractly, comprehend complex ideas, learn quickly and learn from experience. It is not merely book learning, a narrow academic skill, or test-taking smarts. Rather, it reflects a broader and deeper capability for comprehending our surroundings—'catching on,' 'making sense' of things, or 'figuring out' what to do."[7] One of the striking things about this definition is the emphasis on the generality of mental abilities. Another is the phrase "among other things" right at the outset, which implies that even this very broad definition is not exhaustive.

Compare this to a similarly comprehensive definition of creativity that reflects "the capacity of persons to produce compositions, products, or ideas of any sort which are essentially new or novel, and previously unknown to the producer. It can be imaginative activity, or thought synthesis, where the product is not a mere summation. It may involve the forming of new patterns and combinations of information derived from past experience, and the transplanting of old relationships to new situations and may involve the generation of new correlates. It must be purposeful or goal directed, not mere idle fantasy although, it need not have immediate practical application or be a perfect and complete product. It may take the form of an artistic, literary or scientific production or may be of a procedural or methodological nature."[8]

Like intelligence, then, creativity is viewed as a general capacity. But therein ends the similarity. Intelligence is about making sense of what is, creativity is about what might be. Intelligence is mainly about getting things right in the known space and leans toward certainty. Creativity involves venturing into unknown realms and tends toward risk. To draw on the distinctions in types of reasoning as expounded by Charles Sanders Peirce (1839–1914),[9] many assessments of intelligence largely track the ability to engage in deductive reasoning (inference based on premises) and inductive reasoning (inference based on observation). Abductive reasoning (hypothesis generation), on the other hand, is the playground of creativity.

They appear to be quite different abilities, so why are creativity and intelligence talked about so often in the same breath? As it turns out, the answer has to do with the history of the measurement of creativity—which is closely tied with the history of measurement of intelligence.

The Beginnings of Intelligence Testing in a Nutshell

Charles Spearman was the first to provide evidence of intelligence as a general mental ability. He did so by noting high positive correlations between children's performance on very different subjects in school, including the classics, French, English, mathematics, and music, and thereby inferred that a general capacity underpinned performance on all tasks of mental ability. He referred to this as "general intelligence." In his own words: "The above and other analogous observed facts indicate that all branches of intellectual activity have in common one fundamental function (or group of functions), whereas the remaining or specific elements of the activity seem in every case to be wholly different from that in all the others. The relative influence of the general to the specific function varies in the ten departments here investigated from 15:1 to 1:4."[10] Note that Spearman distinguishes between general intelligence, or the "g-factor," and special functions or abilities ("s-factors"). The latter are domain-specific (arithmetic, visuospatial, musical, logical, etc.) and are associated with substantial individual variability. Savant skills, explored in the previous chapter, are high-performance instantiations of s-factors. The need to consider intelligence in terms of a variety of specific instantiations, as opposed to a general factor, was most notably advocated by Howard Gardner in his theory of multiple intelligences.[11]

Over a century of research has largely backed Spearman's theory of general intelligence. Evidence confirms that different mental abilities are not orthogonal, as performance on any one task tends to correlate with performance on another cognitive task. This means that high performance on any one task of mental ability (e.g., word fluency) is associated with the increased probability of high performance on another task of mental ability (e.g., perceptual speed). Logic would dictate that this is only possible if there is an underlying common factor that underpins all mental operations. An examination of the structure of mental abilities can be carried out by assessing the relationships between a range of individual cognitive tasks, the overarching mental abilities they assess (e.g., reasoning, spatial ability, memory, processing speed, vocabulary), and the g-factor. The correlations between such variables at all levels of analysis have been shown to be not only positive, but also extraordinarily high (ranging between 0.74 and 0.96).[12]

Raymond Cattell (1905–1998), whose work has had a lasting impact on how measures of intelligence are designed, is credited with distinguishing between two aspects of general intelligence.[13] Fluid intelligence (Gf) reflects the capacity for reasoning that is minimally dependent on prior learning and "has the character of a purely general ability to discriminate and perceive relations between any fundaments, new or old." In contrast, crystallized intelligence (Gc) is dependent on acquired knowledge and "consists of discriminatory habits long established in a particular field, originally through the operation of fluid ability, but not longer requiring insightful perception for their successful operation."[14] This distinction between general reasoning-based (fluid) intelligence versus knowledge recall-based (crystallized) intelligence is worth bearing in mind, given that intelligence measures that assess fluid intelligence are most commonly employed in assessing the association between creativity and intelligence, and that fluid intelligence overlaps almost perfectly with the g-factor.

The measuring of intellectual abilities began with the efforts of Alfred Binet (1857–1911), who was commissioned by the French government in 1904 to design methods by which it would be possible to distinguish between the learning capacities of children to identify those who would not benefit from standard classroom instruction (owing to their impoverished intellectual capacities) in order to provide them better-suited remedial educational services. Binet and his collaborator Théodore Simon (1873–1961) developed the first test of intellectual ability, which was called the Binet-Simon scale, and emphasized the following: "It seems to us that in intelligence there is a fundamental faculty, the alteration or the lack of which, is of the utmost importance for practical life. This faculty is judgment, otherwise called good sense, practical sense, initiative, the faculty of adapting one's self to circumstances. To judge well, to comprehend well, to reason well, these are the essential activities of intelligence." Binet was also careful to note the limitations of the test he developed in that "this scale properly speaking does not permit the measure of the intelligence, because intellectual qualities are not superposable, and therefore cannot be measured as linear surfaces are measured, but are on the contrary, a classification, a hierarchy among diverse intelligences; and for the necessities of practice this classification is equivalent to a measure."[15]

This measure saw the weighing of a person's mental age in comparison to their chronological age. The latter simply reflects how old a person is based on the number of years that have elapsed since their birth. If a person's intellectual performance was comparable (developmentally aligned), better (developmentally advanced), or worse (developmentally delayed) than that of their chronological age-matched peers, they were classified as average, high, or low in intelligence, respectively. The intelligence quotient or IQ of a person was first calculated by dividing a person's mental age by their chronological age and multiplying that fraction by 100.[16] As IQ is now established to be normally distributed in the population (following a bell-shaped curve),[17] more contemporary intelligence tests calculate IQ on the basis of statistical properties of a normal distribution (and deviations thereof), where the average or mean score is 100 and the standard deviation or spread of scores around the mean is 15 points. An average IQ score falls in the range of 85–115 (i.e., within the zone of the mean, plus or minus one standard deviation) and it typifies 68 percent of the population. Ninety-six percent of the population have a score that falls into the 70–130 IQ range. The 2 percent of the population who obtain scores under two standard deviations (threshold IQ = 70) are considered intellectually impaired, whereas the 2 percent who obtain scores over two standard deviations (threshold IQ = 130) are often classified as intellectually gifted or exceptional.

The importing of the Binet-Simon tests to the United States saw several fundamental changes in the way the scale was refined and employed. The dissemination of the idea that intelligence is fixed and heritable accompanied these revisions. That Binet was resolutely opposed to such views is captured by his impassioned statements delivered two years prior to his early demise: "Some recent philosophers seem to have given their moral approval to these deplorable verdicts that affirm that the intelligence of an individual is a fixed quantity, a quantity that cannot be augmented. We must protest and react against this brutal pessimism; we will try to demonstrate that it is founded on nothing."[18] Binet would have found some solace in knowing that there is evidence to show that IQ scores can be improved with training.[19]

The focus of examination in intelligence testing also completely shifted 180 degrees in the United States when it was purposively deployed in service of the identification of high performers for the first time. The American version of the test, the Stanford-Binet intelligence scale, which was developed by Lewis Terman (1877–1956), was amended to such a degree that the scale was essentially removed from its original psychiatric context and repurposed as a psychological instrument.[20] It even resulted in the removal of the inventive and imaginative tasks that Binet and Simon incorporated in their original scale.[21] Terman was a passionate advocate of intelligence assessment and almost single-handedly oversaw the implementation of IQ testing across the country; first in the context of schools, followed by the screening of army recruits in World War I, and then far beyond. With it came the widespread public acceptance of intelligence testing that continues to the present day.

That Terman's avowed interest was toward the gifted end of the intelligence spectrum is clear from the following statement: "The number of children with very superior ability is approximately as great as the number of feeble-minded. The future welfare of the country hinges, in no small degree, upon the right education of these superior children. Whether civilization moves on and up depends most on the advances made by creative thinking and leaders in science, politics, art, morality, and religion. Moderate ability can follow or imitate, but genius must show the way."[22]

Genius IQ = Genius Creativity

Terman clearly viewed creativity as being rooted in intelligence, and equated genius in intelligence with genius in creativity. In this he followed closely in the tracks of Francis Galton (1822–1911), the polymath who is credited with founding the field of psychometrics, the science of measuring latent human faculties to objective standards. Galton regarded intellectual and creative exceptionalism as emerging from general capacities that were not only measurable, but also heritable. In fact, he went further to also ascribe the drive and motivation to excel as inevitable consequences of such abilities: "A gifted man is often capricious and fickle before he selects his occupation, but when he has been chosen, he devotes himself to it with a truly passionate ardour. . . . He will display an insight into new conditions, and a power of dealing with them with which even his most intimate friends were unprepared to accredit him."[23]

While Galton's views were primarily rooted in the examination of the biographical details of eminent achievers in the past, Terman's focus was on the present and the future. He spearheaded what went on to become not only the first but also the longest longitudinal investigation in the field of psychology to date. The Terman Study of the Gifted, originally called the Genetic Studies of Genius, began in 1921 at Stanford University as a comprehensive examination of the development and maturation of gifted children from early childhood to late adulthood in every respect imaginable—physical health,

personality characteristics, educational attainment, professional achievement, sexual practices, marital success, mortal-
ity, wellbeing, and so on. For around half of the participants
recruited, the process began with nominations from classroom
teachers in grades 1–8 to provide the names of the youngest
child as well as the brightest children in their classes from
the current and previous year. All nominated children were
then subject to different tests of intellectual ability according
to their age group. For the remaining half of the participants
recruited, other testing protocols were observed. The general
aim was to identify at least a thousand children who were in
the highest 1 percent in the school population of California in
terms of their general intelligence. The original group included
1,528 participants who were assessed as having an IQ score of
135 or more.[24] The "Termites" as they came to be known were
not only extensively tested upon study recruitment, but also
assessed in more than ten follow-up studies that took place
predominantly every five years over 1928–1960 and 1972–
1986 and continued more sporadically into the early 2000s.[25]

The findings indicated that the Termites showed higher
educational and professional attainment on average than the
general population. Scholarly achievement in this gifted sample was far higher than the California average at that time,
with over 80 percent of them going to college, and over 60
percent going to graduate school. More than a hundred patents had been awarded to three of the Termites. There were
fifty-two doctors and eighty-two lawyers among the sample.
Their literary achievements were also considerable, with
around ninety books and 1,500 articles published. The group
as a whole earned much more money than the average person.

Nonetheless, these outcomes were well short of what Terman had hoped for. The attained grades, for instance, were

only moderately high with less than 15 percent of the Termites achieving an A average. With regard to the occupational groups to which they belonged, while around 48 percent were classified as professionals, the remaining 52 percent were not. Instead, they pursued occupations that were classified as falling in the semi-professional and managerial category, the clerical, skilled, and retail category, and the agriculture category. Especially notable were the absences in expected exceptional achievement within this gifted group. No one from the group founded an industry or became an artist or a leader of renown. The lack of Nobel Prize winners, for instance, was an especially awkward point of note, as two people who went on to win the .Nobel Prize in science—William Shockley and Luis Alvarez— were tested for inclusion in the Terman study but not selected.

Even Terman admitted the following in this twenty-five-year follow-up of his gifted sample in 1947: "At any rate, we have seen that intellect and achievement are far from perfectly correlated. . . . So little do we know about our available supply for potential genius, the environmental factors that favor or hinder its expression, the emotional compulsions that give it dynamic quality, or the personality distortions that make it dangerous."[26] The words of Sarah Ann Albright mentioned at the start of this chapter in many ways gets to the heart of what was ignored in the Terman studies. The endowment for intellectual achievement does not necessarily translate to the determination and drive required for great accomplishments or the inclination toward creative engagement.

Genius IQ ≠ Genius Creativity

The 1950s saw the emergence of the tide against the simple equation of intelligence (particularly as measured by IQ tests) and creativity. Far from being a niche view or one bellowed from the margins, eminent scholars in the field of individual differences testing recognized the limitations in this logic.[27] Several viewed intelligence and creativity as being quite separable. Louis Thurstone, for instance, noted that students of higher intelligence do not necessarily generate the most creative ideas, and how it was a rarity to come across individuals who have the combination of being both highly intelligent and highly creative.[28] Frank Barron also speculated on the differences between the two: "To use another of the distinctions proposed by Freud in his theory of the functioning of the mental apparatus . . . primary process thinking to the exclusion of the secondary process marks the original but unintelligent person, secondary process thinking which carries ego-control to the point where the ego is not so much strong as muscle-bound marks the intelligent but unoriginal person, and easy accessibility of both primary process and secondary process marks the person who is both original and intelligent."[29]

The need to consider the different forms of giftedness was expounded in the highly influential and widely debated work of Jacob Getzels and Philip Jackson.[30] They openly made a case for the consideration that one could be gifted not only in terms of intellectual ability, but also in other dimensions like creativity, psychological adjustment, and morality. This idea was tested by examining select groups of cognitively gifted children from a sample of around four hundred and fifty adolescents. These young individuals exhibited cognitive excellence on either a single IQ measure or across five creativity measures. The gifted IQ group included 28 children who were high in intelligence (in the top 20 percent on IQ) but not in creativity (below the top 20 percent on creativity). The gifted creativity group included 26 children who were high in creativity (in the top 20 percent on creativity) but not in intelligence (below the top 20 percent on IQ). These two groups were subject to a series of examinations on school performance, need for achievement, teacher perception, values, fantasies, career aspirations, family circumstances, and so on.

They found that both the gifted IQ and gifted creative groups were undifferentiated in relation to school achievement with both groups showing high performance. On tasks that necessitated the use of fantasy, the gifted creative group outperformed the gifted IQ group in showing more unexpected endings and engaging in stimulus-free themes as well as in the use of incongruities, humor, playfulness, and violence in their generated stories. While teachers rated the gifted IQ children to be more desirable as students than the average adolescent, a similar positive estimation bias was not found for gifted creative students—this is a noteworthy point especially when considering that key studies on intelligence and achievement (such as Terman's) relied on teacher nominations for study inclusion.

Both groups were also comparable in their motivation for achievement; however, they differed greatly in whether they felt that the personal traits they valued for themselves would lead to success in adulthood. While the gifted IQ group endorsed that association, the gifted creative group did not agree that the personal attributes they valued for themselves would lead to future success in their lives. A key finding of this study was low correlation between creativity measures and IQ, a pattern that was even more pronounced in the lab-based investigations of Michael Wallach and Nathan Kogan on younger children.[31] In concluding that their gifted IQ students showed greater engagement in 'convergent' modes of thinking, whereas the gifted creative students displayed more 'divergent' modes of thinking, Getzels and Jackson drew on concepts from the seminal work of Joy Paul Guilford, who is credited with launching the scientific study of creativity from the sidelines to the spotlight.

In his famous 1950 address at Pennsylvania State College as the president of the American Psychological Association, Guilford averred that "creativity and creative productivity extend well beyond the domain of intelligence."[32] In fact, his structure-of-intellect (SOI) model espoused a very broad view of intelligence that included mental abilities of relevance to creativity. Guilford saw "intelligence as a collection of abilities or functions for processing information."[33] The instantiation of these abilities was dependent on the type of informational *content* (visual-figural, auditory-figural, symbolic, semantic, behavioral), the *products* of this information (units, classes, relations, systems, transformations, implications), and the intellectual *operations* performed with this information (memory, cognition, evaluation, convergent production, divergent production).

While he acknowledged the role of all the intellectual operations in the context of creativity, Guilford singled out divergent production—which occurs when one is faced with an open-ended problem to which a broad search is applied, and for which there are several possible answers—as being central to creativity. Guilford adopted a factor-analytic approach to derive the common variables that explained task performance across different measures of divergent thinking. For instance, in 1952, using factorial analysis of the performance of 301 air cadets and 109 student officers on fifty-three tests, he found evidence for nine creativity-specific factors.[34] These were identified as sensitivity to problems, originality, redefinition, closure, fluency (three types: associational, ideational, word), and flexibility (two types: adaptive, spontaneous). A further factor, elaboration, was added in later studies.[35] Creativity measures have by and large been developed with these factors in mind ever since.[36] Guilford himself saw the distilling of these components into four essential divergent thinking abilities (fluency, originality, flexibility, elaboration) that could be captured by a single test—the alternate uses task—that was published in 1960.[37] Even today, the alternate uses task remains among the most widely used measures in the field of creativity research.

Guilford's ideas received criticism not so much for his assertions on the intelligence-creativity association, but instead for facets specific to his model,[38] such as low correlations between the different measures of creativity. For instance, Robert Thorndike, the educational psychologist and psychometrician, accepted that divergent thinking is likely to be distinct from what is captured by conventional IQ tests that measured the general factor of intelligence. He noted, however, that the evidence did not actually confirm the presence of a general factor of creativity as would be expected by Guilford's model because correlations between performance on different divergent thinking tasks were generally low: "In particular, we should ask whether there is any common characteristic running through these tests to which the common term 'creativity' may legitimately be applied. This is an important question, because if the various so-called 'creativity tests' are measuring different and largely unrelated characteristics of people, then using a common term to include all of them, or applying a common designation to groups of people identified by different ones of them, can be productive of nothing but confusion."[39]

The effort to model creativity testing after intelligence testing laid bare the significant limitations of the approach. Although the field of intelligence measurement is generally deemed to be highly psychometrically sound, IQ testing is roundly criticized for not capturing much of what people intuitively understand to be intelligence. This is in spite of the sizeable evidence that shows stable and expected patterns of a higher probability of more positive outcomes in professional and personal domains in association with higher levels of intelligence as measured by IQ tests.[40] Despite the considerably worse problems in creativity assessment in relation to both test validity (the extent to which the measures do, in fact, assess the construct being tested) and test reliability (the consistency of the measures in assessing the construct), the level of the criticism directed toward creativity research methods has been paltry in comparison. Indeed, even when they are put forward,[41] the critiques are largely ignored. Both the assessment of creativity using these tests as well as the examination of the creativity-intelligence link continues unabated.

The Threshold Theory

Several postulations have been put forward to characterize the creativity-intelligence link: that creativity is a subset of intelligence, that intelligence is a subset of creativity, that intelligence and creativity are overlapping sets, that creativity and intelligence are nonoverlapping or disjointed sets, and that creativity and intelligence are coincident sets such that creativity is the pinnacle of intelligent thought and action.[42] Of all the ideas in play, the intelligence and creativity as overlapping sets notion and, in particular, the "threshold hypothesis" or "threshold theory" have been recipient of the most concerted empirical attention.

In 1953, John Drevdahl sought to empirically assess the associations between creativity and a range of personality and intellectual factors in a group of university students. They were divided into creative and noncreative groups based on the judgment of two raters who evaluated each participant on a seven-point scale of creativity based on the raters' experience of the person. While the creative group produced a higher proportion of original responses on a creative thinking task and showed higher scores on personality variables such as radicalism and self-sufficiency, they were indistinguishable from the noncreative group in assessments of intellectual ability. The author's conclusion of such findings reads as follows: "Although there does not appear to be a significant relationship between creativity and general intelligence in this population, it would be erroneous to assume that general intelligence or the other nonsignificant intellectual factors studied are unimportant. It seems more likely that these factors may be necessary but insufficient characteristics for creativity."[43]

This assertion is interesting for two reasons. First, it emphasizes a relationship between creativity and intelligence despite finding no evidence for the same. This is not an uncommon occurrence in the literature—the insistence that there is a singular association to be found between creativity and intelligence even when the evidence offered up within the study in question does not support it or only attests to a weak association. Second, it stipulates a conditional relationship—necessary but insufficient—between intelligence and creativity.[44] This view holds that one needs some degree of intelligence to be creative, but more intelligence does not translate to higher creativity beyond a given point. The big question that has stimulated an avalanche of research is: What is the threshold or cut-off IQ value beyond which intelligence does not predict creativity? The standard view that has been around since the time of the Getzels and Jackson 1962 study is that the cut-off point corresponds to an IQ score of 120.[45] The central idea of the threshold theory is that intelligence and creativity are positively correlated up to an IQ level of 120, but beyond that there is no significant association between the two.

On balance, the accumulated evidence over more than half a century does not by any estimation lend unequivocal support to the threshold hypothesis. Scholars in the field, however, fall into one of two camps when coming to terms with the overall picture. There are those, on the one hand, who point to evidence showing the negligible relationship between intelligence and creativity,[46] and call for the outright rejection of the threshold theory.[47] On the other hand are those who attribute the lack of evidence in favor of the threshold hypothesis to be rooted in factors specific to study design and data analytical approaches.[48] The basic underlying idea here is that greater consistency across studies in relation to such factors—how they are conceived, tested, and examined—would enable the stable capture of findings that would not only support the threshold hypothesis, but also help refine it.

An example of such a refinement of the hypothesis was the finding that the threshold actually differs depending on the creativity measure in question. The measure of ideational fluency on the alternate uses task reflects the total number of ideas generated. This measure was associated with an IQ threshold of 86. What this means is that IQ and creative potential (as assessed by ideational fluency) were positively correlated up to an IQ of 86, but not beyond that point. A much higher IQ threshold of 120 was found for the average originality measure on the alternate uses task, which reflects the uniqueness of the generated ideas. Therefore, IQ and creative potential (as assessed by ideational originality) were positively correlated up to an IQ of 120, but not beyond that point.[49]

The big picture of this field of research reveals that the results are often critically swayed by a range of factors that are as elemental to individual differences as sociodemographic variables (e.g., age, gender, profession) and as complex to the scientific method as data analysis choices (e.g., assessing raw versus standard scores, using segmented versus quantile regression). Even slight differences can lead to contradictory findings. The one thing that can therefore be dependably gleaned is that the evidence certainly does not support a general threshold-type relationship between creativity and intelligence. The threshold hypothesis, if it does apply, is elusive and highly specific.

Concluding Thoughts

Trying to uncover the precise nature of the association between creativity and intelligence has proved exceedingly tricky because of the many factors that prevent a clear picture from being drawn. The correlations between the two vary widely depending on the population being studied,[50] the type of creativity measure being used,[51] and the manner in which the creativity task has been scored, with studies using objective scoring protocols showing lower correlations with intelligence measures than those using subjective scoring protocols.[52]

Contemporary scholars nonetheless continue to advance this field of study by following alternative paths when considering the relationship between creativity and intelligence. Some have zoned in on the minutiae of intelligence testing by drawing the attention away from fluid intelligence (Gf) and emphasizing other aspects of general intelligence like crystallized intelligence (Gc)[53] and broad knowledge retrieval abilities (Gr) that take the form of long-term memory access and storage.[54] Others have taken the route toward more specific forms of intelligence and have shown that higher levels of specific intelligences are indeed associated with higher levels of domain-specific creativity. For instance, two cohorts of mathematically precocious thirteen-year-olds who were identified in the 1970s as being in the top 1 percent in mathematically reasoning abilities were followed up four decades later and found to have made significantly higher scientific creative contributions than the national average in the form of scientific publications, books, patents, and so on.[55] That an association between creativity and intelligence may be strongest at the highest IQ range, a suprathreshold hypothesis of sorts, was formulated as early as 1957 by Frank Barron: "It is of course reasonable to expect that intelligence and originality will covary positively. If one defines originality as the ability to respond to stimulus situations in a manner which is both adaptive and unusual, and if one defines intelligence simply as the ability to solve problems, then at the upper levels of problem-solving ability the manifestation of intelligence will be also a manifestation of originality. That is to say, the very difficult problem which is rarely solved requires by definition a solution which is original."[56]

Others have taken an altogether different direction by formulating wider theoretical models that take into consideration other factors, such as personality and contextual variables, in making sense of how creativity and intelligence work together. Prominent examples include the amusement park theory, investment theory, triangular theory, and the theory of successful intelligence.[57] These largely remain empirically untested theoretical ideas, owing to the difficulty in designing studies that provide adequate examinations of such models. Nonetheless, they are notable for drawing attention to the role played by *other* factors in the equation that derive from outside the individual (e.g., sociocultural environment) or emanate from within the individual (e.g., creative motivation).

The startlingly limited views of highly esteemed and influential figures like Galton and Terman on the impact of "nurture" on the creativity-intelligence link is amply recorded in their best-known works. Galton went so far as to deny any role of the environment in long-term achievement. To him, the gifted were granted not only intelligence but also an indefatigable spirit that would see them through any circumstance and propel them to their inevitable destiny of greatness. Terman also believed that intelligence was entirely accounted for by innate factors, but he did see the importance of nurture after the fact—that fitting learning environments for the gifted (by, for instance, promoting them more rapidly) were necessary to allow the fruition of the full extent of their innate intellectual capital. Both were among the most erudite of men, with deep respect for the scientific method and extensive expertise in quantitative methodology. But their own phenomenal intellectual abilities did not serve as enough of an aid in overcoming the blind spot that arose from ignoring the extensive and tremendous intellectual and creative achievements of "ordinary people" whose IQ scores fall outside the top 2 percent.

The singular role of not only opportunity but also chance in creative achievement is eloquently captured in these reflections of W. E. B. Du Bois: "We have, to be sure, a few recognized and successful Negro artists; but they are not all those fit to survive or even a good minority. They are but the remnants of that ability and genius among us whom the accidents of education and opportunity have raised on the tidal waves of chance. We black folk are not altogether peculiar in this. After all, in the world at large it is only the accident, the remnant, that gets the chance to make the most of itself but if this is true of the white world it is infinitely more true of the colored world."[58]

Internal factors like interest and drive are also rarely given their due. In describing interest as "the feeling of enthusiasm, mental excitation, and attraction to a subject," Edward Thorndike (1874–1949) emphasized an essential feature of "interest," namely that it has an inherent driving force that leads to a striving to "devote one's thoughts and actions to some kind of phenomenon."[59] Factors related to interest like openness to experience and intrinsic motivation have received abundant attention,[60] but the drive to engage in creative pursuits has been mostly overlooked.[61] This is surprising, especially as the need to focus on such factors was noted well over a century ago by Theodule Armand Ribot (1839–1916), who proposed one of the earliest theoretical frameworks for the creative imagination: "This group of facts shows us the existence, beyond images, of another factor, instinctive or emotional in form . . . which will lead us to the ultimate source of the creative imagination."[62]

In his book *The Story of Measurement*, Andrew Robinson provides a convincing overview of the ubiquity of measurement in all aspects of human life with the bittersweet reflection that, while "measurement is necessary for civilization to flourish," "it reduces human values to inhuman numbers" and this conflict explains our simultaneous attraction toward and rebellion against quantification.[63] Scholars continue to zone in on intelligence testing in the context of creativity not because it is the most important or relevant factor to consider. We do so on historical grounds. Creativity measurement has its roots in intelligence testing and is thereby bound to it in a foundational manner. An inability to move past this state of affairs is to be expected, as paradigm shifts are rare occurrences in any domain of scientific study.[64] It is easier, after all, to flow with rather than against the tide. But there is another reason why scholars stick with intelligence testing in their examination of the creative mind. We do so because intelligence is the simplest factor to measure to an objective and psychometrically sound standard. This is not true of most other variables of interest to creativity, and not even of the measurement of creativity itself. The appeal of quantification lies in the security it bestows on the trustworthiness of the conclusions we draw about any phenomena under study. But it would be unwise for us to disregard its deep hidden pitfalls. The reflections of Theodore Porter (1995) aptly capture this point: "Quantification is a way of making decisions without seeming to decide. Objectivity lends authority to officials who have very little of their own."[65]

6 Dopamine and Creativity

There had been an account in early 1967 of the effects of L-DOPA on people with ordinary Parkinson's disease, and it was touted as a sort of miracle drug. Now one of my patients—in fact, the original Leonard L., who was somewhat different from the movie version, a very bright man—he was the one who drew my attention to this, and he spoke of dopamine, the neurotransmitter needed in the brain, as "resurrectamine." He spoke of Cotzias, the physician who had introduced L-DOPA, as "the chemical messiah." You know, one sees the depth of intelligence and hope and irony and desperation here.
—Oliver Sacks[1]

"To think is to move," and this control is DA[dopamine]-dependent.
—Keeler, Pretsell, and Robbins[2]

Just as apparent creativity in nature can be understood with reference to the elements of natural selection, apparent human creativity may be best understood with reference to the elements of reward-related incentive learning.
—Richard J. Benninger[3]

Every once in a while we have a "gut feeling" about a choice that does not have an easy explanation—it was the dopamine that made us do it.
—Terence Sejnowski[4]

Dopamine is arguably the most renowned of all brain chemicals. With its celebrity-like status, the word "dopamine" is indiscriminately bandied about across a range of disparate contexts, from corporate settings and high schools to yoga studios and self-help books. It is typically touted as the magic substance that lies within us that has the potential to help us enhance our performance or cultivate the ideal mindset that leads to grand successes.[5] That dopamine is synonymous with pleasure is taken to be a general and important truth. After all, we need to find pleasure in our productive daily activities in order to function optimally, yet too much pleasure from the unproductive pursuits can lead to our downfall. This is why New Age gurus, modern-day influencers, scientific experts on the public engagement front, and countless others (who often mean well) can glibly encourage a public that is eager for easy answers to regard dopamine as both desirous and dangerous, while promoting the idea that the ability to channel its powers is within our grasp.[6]

Let us reflect on these popular narratives surrounding the power of dopamine against the facts on hand. Dopamine is one of more than one hundred known neurotransmitters.[7] Neurotransmitters are generally defined as substances that are synthesized and stored within neurons, and they work as chemical messengers in facilitating communication between neurons as well as other cells (glia, muscles) in the central nervous system via the gap between the cells (the synapse). Upon the firing of an action potential in a neuron, its neurotransmitters are released into the synapse and these chemicals then act on adjacent neurons and other postsynaptic cells to cause excitation or inhibition of those cells. Estimations of the percentage of synapses in the brain that are accounted for by each neurotransmitter indicate that dopamine neurons account for around 1 percent of the total.[8]

This low percentage, however, does not capture the widespread impact of these neurons on brain function. This is because dopamine is neuromodulatory by nature, and neuromodulators use volume transmission during intercellular communication. "Volume transmission" refers to a diffuse release of neurotransmitters that renders it possible for neuromodulators to tune brain circuitry well beyond individual neurons (where signal transmission is typically limited to the synaptic cleft: "synaptic transmission") to expand across the extracellular space and thereby influence whole brain regions. Neuromodulators work by changing the synaptic or cellular properties of specific neurons and thereby alter the neurotransmission properties of those cells. Dopamine in particular is characterized by rapid and localized release, followed by diffuse volume transmission. The signaling patterns of dopaminergic neurons are fundamentally characterized by dynamic switching between tonic and phasic release. Tonic release maintains steady-state concentrations of dopamine and thereby the normal functioning of neural circuits. In contrast, phasic release is generated by burst firing which produces brief and rapid dopamine transients in downstream structures that lasts for many seconds. What's more, dopamine also co-releases other neurotransmitters like glutamate (excitatory) and GABA (inhibitory).[9]

Dopamine therefore wields its influence in the brain by dynamically sharpening or weakening the communication between certain neurons. To take an example: the transmission of sensory information from the cortex to the striatum (a brain structure that facilitates voluntary movement and reinforcement learning, among other functions) is modulated by dopamine. Dopamine works by preferentially highlighting the inputs that are the most significant at that moment in time to the organism in question (e.g., food to a hungry mouse), and it does so by strengthening highly relevant excitatory cortical input (e.g., the smell of food) and suppressing less relevant cortical excitatory inputs (e.g., the sound of birds singing in the trees). In the absence of dopamine, all sensory inputs from the cortex are treated as equal and the mouse shows apathy, seemingly indicating that it does not know what to do and therefore does not respond at all.[10]

The functions modulated by dopaminergic neurons are typically tied to its specific brain pathways. Most dopamine neurons within the mammalian brain originate from two nuclei in the ventral midbrain—the pars compacta of the substantia nigra and the ventral tegmental area—and from there project far and wide beyond the midbrain. The dopamine neurons radiating from the substantia nigra project to the dorsal striatum (nigrostriatal pathway), whereas those emanating from the ventral tegmental area mainly project to the ventral striatum, the prefrontal cortex, and limbic regions like the hippocampus (mesocorticolimbic pathway). The nigrostriatal pathway is regarded to be mainly involved in facilitating motor functions, such as action control and action selection, whereas the mesocorticolimbic pathway is held to facilitate reward processing and reinforcement learning. However, mounting evidence indicates fundamentally intricate interplay between these pathways in complex aspects of cognition and behavior. The fact that both nigrostriatal and mesocorticolimbic pathways are implicated in severe disorders of psychological function, such as addiction and schizophrenia, certainly suggest this is the case.[11]

So how does creativity come into this picture? Let us explore the role of dopamine in the context of creative thinking, with reference to each of the dominant views of its essential function.

Dopamine the Rewarder

The first clear-cut evidence of a reward center in the brain came from the extraordinary work of James Olds and Peter Milner in the 1950s, which demonstrated that when rats with electrodes implanted deep in their brains, particularly in the septal area, were given the opportunity to press a lever at will in order to self-stimulate their own brains, they did so repeatedly and regularly over extended periods of time.[12] Electrodes implanted in sensory and motor systems were associated with self-stimulation response rates at the level of chance (ten to twenty-five lever presses per hour), whereas those in specific parts of the hypothalamus and midbrain areas, known to be involved in the control of sexual, digestive, and excretory processes among others, showed response rates of 500 to 5,000 presses per hour. This clearly demonstrated that the electrical stimulation delivered in these brain regions was experienced as rewarding by the rats. In fact, in some of these areas, electrical stimulation was found to be far more rewarding than a primary reinforcer like food. Hungry rats ignored food that was readily available to them and chose the option of self-stimulation instead, with some rats doing so "more than 2,000 times per hour for 24 consecutive hours!"[13]

These seminal discoveries gave rise to concerted research on the reward and motivation systems of the brain across different species. Reward-specific pathways were identified that spanned the forebrain (e.g., frontal lobe, basal ganglia), the midbrain (e.g., ventral tegmentum, substantia nigra), and the hindbrain (e.g., brain stem, cerebellum), and the key role played by dopamine in facilitating positive reinforcement or reward-based behavior became increasingly recognized.[14] We know now that the excitation of dopamine neurons in the ventral tegmental area is associated with all addictive drugs. Indeed, the production of high levels of dopamine following the ingestion of such drugs leads to the engagement of the brain's reward system in facilitating the addiction cycle.[15]

That dopamine helps facilitate creativity is indirectly derived from evidence showing its involvement in the experience of pleasure.[16] Modest improvements in creative performance can result from inducing positive moods[17] and enhanced cognitive flexibility has been shown to follow mild increases in positive affect.[18] For instance, across a series of four experiments, only the induction of positive emotions via various means (e.g., watching a comedy film; receiving a small gift of candy) improved consequent creative performance. This effect did not result with negative emotions (e.g., watching a film segment about Nazi concentration camps) or when engaging in merely physiological arousal-inducing tasks (e.g., physical exercise) or neutral activities (e.g., watching a film segment about math).[19] However, this improvement in cognitive performance following the induction of positive affect is also demonstrated in the performance of tasks that do not necessitate creative thinking.[20] Positive affect is thereby posited to improve cognitive flexibility in general,[21] as opposed to being particular to creativity.

What's also interesting to note in this context is that the nature of the rewarding experience also plays a role. The impact of expected extrinsic rewards that are contingent on task engagement and performance are in fact widely regarded as being detrimental to creativity when it is examined outside lab settings (e.g., in real-world contexts like the workplace) when contrasted with intrinsic rewards. A rewarding experience that stems from intrinsic motivation "or the drive to do something for the sheer enjoyment, interest, and personal

challenge of the task itself" is considered to be not only conducive but also pivotal to creativity.[22]

In any event, a narrow focus on reward-facilitated cognition vis-à-vis the role of dopamine leads to overlooking the
larger picture. This is because it is now well established within
the neuroscientific literature that dopamine neurons do not
solely respond as a function of rewards or reward-related information.[23] Instead, they additionally code for uncertainties
in the form of reward prediction errors that reflect whether
the transpired experience is better or worse than what was
expected.[24] Moreover, they are also sensitive to other motivationally relevant occurrences, such as novel events,[25] aversive
outcomes (which are the diametric opposite of rewards),[26]
and high-intensity events that are nonappetitive (not relating to food, drink, or copulation).[27] Extensive neurochemical
and electrophysiological evidence indicates that the highly
differentiated dynamics of dopamine neurotransmission (in
terms of both temporal activity and volume concentrations)
enables the communication of "differential and heterogenous
information to subcortical and cortical brain structures about
essential outcome components for approach behavior, learning and economic decision-making."[28] Equating dopamine
solely with rewards or pleasure, therefore, constitutes a misattribution of its core underlying functions.

Dopamine the Controller

Cognitive control or executive functions refer to the capacity to flexibly fine-tune and regulate behavioral responses to
carry out goal-directed and intelligent thought and action.
Key processes include working memory, inhibition, attentional control, set shifting, stimulus selection, planning, and

decision-making. The prefrontal cortex sits atop the hierarchy of brain regions in the frontostriatal pathways that orchestrate cognitive control within which dopamine plays a central role.[29] Specifically, dopamine influences cognitive control in the prefrontal cortex by facilitating three functions: the gating of sensory input, the maintenance and manipulation of information in working memory, and the transmission of motor commands.[30] Striatal dopamine, in contrast, plays a key role in cost-benefit decision-making in this context. It influences cognitive control by augmenting the sensitivity to the benefits of cognitive effort, while concurrently diminishing the sensitivity to its costs.[31]

An influential theory that draws on the relevance of the control hypothesis of dopamine on creativity posits that creativity benefits from moderate levels of dopamine (inverted U-function: not too high and not too low) across both of its pathways. Moderate amounts of dopamine in the striatum (nigrostriatal pathway) promotes ideal levels of flexibility-driven creativity, with too little leading to rigidity and too much resulting in distractibility. Moderate levels of dopamine in the prefrontal cortex (mesocortical pathway), on the other hand, are associated with ideal levels of persistence-driven creativity, with too little leading to distractibility and too much resulting in perseveration or repetitiveness.[32] An example of research that lends some support to this theory is one that uses spontaneous eye-blink rate, a clinical marker of striatal dopamine.[33] Spontaneous eye-blink rate was found to predict divergent and convergent creative thinking in different ways, such that higher levels of striatal dopamine led to impoverished performance on convergent creative-thinking tasks, whereas only medium levels of striatal creativity led to more flexibility during divergent thinking.[34]

However, this theoretical model of dopamine function in creativity does not quite align with the extensive neuroscientific literature that illustrates that the mesocortical dopamine pathway is associated not only with cognitive persistence but also with cognitive flexibility. The involvement of dopamine-relevant activity in the dorsolateral prefrontal cortex is, after all, associated with cognitive set-shifting, updating mental set during working memory, and risk-reward decision-making.[35] The action of prefrontal dopamine through D1-like and D2-like receptors is highly complex. One view holds that they serve seemingly opposing functions, with the "D1-dominated state characterized by a high energy barrier among different network patterns that favors robust online maintenance of information and a D2-dominated state characterized by a low energy barrier that is beneficial for flexible and fast switching among representational states."[36] More recent accounts argue for the inverted-U-type function with respect to the tuning of dopamine D1 receptors and cognitive performance, such that optimal levels of dopamine improve signal-to-noise ratio and enhance the efficiency of prefrontal cortex networks, whereas a surfeit or a lack of dopamine disrupts efficiency but lends more flexibility to the networks in the processing of information.[37]

Direct examinations of the effect of dopamine on creativity have also been less than conclusive. For instance, the ingestion of tyrosine, a biochemical precursor of dopamine, did not lead to improvement in flexibility or any other aspect of divergent thinking.[38] Similarly, the ingestion of a dopamine agonist did not result in improved performance on an anagram task, which suggests that this type of cognitive flexibility is not necessarily mediated by the dopaminergic system.[39]

Studies on people with attention-deficit hyperactivity disorder (ADHD) are also regarded as relevant in this context.

This is because ADHD is a condition that is characterized by disinhibition and distractibility as the primary symptoms, traits that are associated with enhanced creativity.[40] The most common medication prescribed in the treatment of ADHD is methylphenidate, which acts on the dopaminergic and noradrenergic neurotransmitter systems. There is, in fact, evidence for enhanced divergent creative thinking, particularly with low levels of ADHD symptoms,[41] but how much this advantage in creative cognition can be attributed to dopamine specifically is difficult to estimate. After all, evidence that examines the impact of ADHD medications on creative performance can just as easily be used to make a case for other neurotransmitters that are often jointly implicated with dopamine.[42]

More recent views see dopamine as facilitating the motivation underlying cognitive control. This means that it modulates the willingness to exert control rather than the ability to do so.[43] It has even been characterized as serving a "double duty" in "translating incentive information into cognitive motivation."[44] So here again, the motivational aspect of dopamine function is drawn upon when explaining its role in facilitating cognitive control.

Dopamine the Driver

The leading theorist in the characterization of the dopamine-creativity link in relation to motivational drives is Alice Flaherty. By integrating insights from the literature on the role of dopamine in novelty seeking, goal-driven approach motivation, mental imagery, attentional inhibition, and reward processing, she proposes that dopamine essentially facilitates the drive to create.[45]

This focus on creative motivation as a drive is distinct from other ways in which we usually think about motivation in the context of creativity—where the focus has been largely on the influence of intrinsic motivation (performing an activity for its own sake, as it is personally rewarding to do so) versus extrinsic motivation (when one is incentivized to perform an activity—to gain a reward or avoid punishment). The drive to create, in contrast, refers to the agency or impetus that leads to the generation of ideas. Higher levels of dopamine result in an increased drive to create and consequently enhanced idea generation. While a heightened drive to create may not necessarily lead to a higher quality of ideas, mere statistical probability ensures that the higher the number of ideas generated, the higher the likelihood of creative ideas occurring within that ideation set.[46]

That dopamine shows general effects on the ideation process by eliciting higher fluency in idea generation, and that this production of more ideas translates to more specific effects of enhancing uniqueness at each level of fluency (from low to high) is supported by some neuropharmacological evidence. Using an option generation motor task, where the aim was to draw as many paths as possible between two points and the metrics assessed were fluency (how many paths were drawn) and uniqueness (rarity of the drawn paths), healthy individuals exhibited a trade-off between uniqueness and fluency, in that they generated many similar path options or few unique path options. Ingestion of a dopamine agonist led to an increase in overall fluency but a reduction in uniqueness. However, after controlling for the uniqueness-fluency trade-off, dopamine was found to improve uniqueness for any specific level of fluency. The same pattern was also found when examining the performance of patients with Parkinson's disease when they were *on* versus *off* their dopaminergic medication.[47]

The ingestion of medications like dopamine agonists and levodopa is sometimes accompanied by an increase in engagement in artistic activities in art-naïve patients.[48] These are standard treatments for disorders marked by insufficiencies in dopamine, such as Parkinson's disease (PD) and related disorders like restless legs syndrome. However, a study that compared three groups of PD patients with two unaffected control groups revealed that dopaminergic treatment does not necessarily lead to better divergent thinking across every individual.[49] All three PD groups received dopaminergic treatment and included patients who were professional artists prior to the onset of PD (Group A), patients who showed artistic-like output post-PD onset following dopaminergic treatment (Group B), and patients who did not exhibit such behaviors (Group C). The two control groups without PD included professional artists (Group D) and nonartists (Group E). The highest level of overall divergent creative thinking was seen in professional artists with and without Parkinson's disease (Group A and Group D). The patients who showed artistic-like output following dopaminergic treatment post-PD onset (Group B) were second in line as they performed better than the remaining two groups (Group C and Group E) on specific aspects of divergent thinking, such as the ability to generate original and elaborate responses. Nonetheless, the fact that the patients who did not show any artistic-like output following dopaminergic treatment post-PD onset (Group C) did not perform better than Group E suggests that the mere exposure of dopamine treatments does not lead to an emergence of artistic talent or improvement in divergent thinking skills across all individuals. In a sizeable subset, though, dopamine seems to augment the drive to engage in creative pursuits.[50]

The intake of dopamine agonists is not only sometimes associated with a release of artistic tendencies, but also the development of impulse control disorders such as hypersexuality, pathological gambling, compulsive shopping, and eating behaviors in Parkinson's disease.[51] Relatedly, while Tourette's syndrome, which is associated with an excess of dopamine,[52] has long been anecdotally associated with creative impulses,[53] the evidence on hand for the Tourette's-creativity link is mixed at best.[54]

This less-than-straightforward picture may be best understood by taking into consideration the notion that the creative drive, like all motivational drives, follows homeostatic principles. The well-established Yerkes-Dodson Law of the inverted U-function between arousal and performance can be transposed to the context of creative idea generation. "Just as both very low and very high arousal hinder the ability to successfully complete a task, both disinterest and an overintense desire to do creative work can inhibit the open-ended exploration needed for creativity."[55] When people are at peak levels of creativity, they are at the Goldilocks tipping point of creative drive (not too much and not too little, but just right) that facilitates optimal generativity. An additional boost to dopamine that may benefit an averagely creative person will often turn out to be counterproductive for the highly creative because it pushes them into the "too much" zone. There are several strands of research that support the necessity to consider baseline creativity or related personality traits and emotional states relevant to creativity. For instance, the ingestion of a dopamine agonist increased response fluency (more paths) and uniqueness (greater diversity of paths indicating more exploratory behavior) for a given fluency level during an option generation task in low motivated or apathetic individuals, whereas in highly motivated individuals it increased fluency without improving uniqueness.[56]

Similarly, in a double-blind study that examined performance on a range of creativity tasks in healthy participants following the consumption of methylphenidate compared to a placebo, novelty-seeking was generally positively associated with all measures of a verbal creativity task under placebo conditions. The ingestion of methylphenidate was found to differentially impact creative performance as a function of differences in novelty-seeking traits across participants. It improved creative performance (more responses, greater flexibility in responses, higher originality of responses) in people with low levels of novelty-seeking traits. However, it had the opposite effect on people with high levels novelty-seeking traits, as it impaired their creative performance.[57] This corresponds to other evidence of the impact of purportedly neuroenhancing drugs on cognitive performance. When these drugs are found to improve creative performance, the effect is limited to people with low levels of baseline creativity. The effects of these drugs are often counterproductive in people with high-baseline creativity.[58]

Noting the role of multiple neurotransmitter systems on creativity, Flaherty also proposed the two-axis model of the creative drive,[59] with one axis for goal-directedness and the other for arousal level. In this model, the noradrenergic system facilitates the arousal axis (high levels = excitability; low levels = relaxation), whereas the goal-directedness axis is mediated by two other systems.[60] Approach motivation in goal-directedness is orchestrated by the dopaminergic system (low levels = apathy; high levels = addiction) while avoidance motivation in goal-directedness is subserved by the seroto-nergic system (low levels = fear; high levels = indifference). On the one hand, through the release of endorphins, the achieving of goals leads to the experience of pleasure during approach motivation (dopamine) and relief during avoid-ance motivation (serotonin). The failure to achieve goals, on the other hand, results in the experience of agitation in approach motivation and anxiety in avoidance motivation. Both approach and avoidance motivation can lead to creative outcomes depending on the circumstances, but the approach motivation state is generally more conducive to the idea gen-eration process.

Dopamine the Explorer

Approach motivation is not only central to the drive to cre-ate. It is also closely tied to another dominant focus in the dopamine-creativity link, namely the drive to explore. Dopa-mine is associated with exploration-relevant traits such as novelty-seeking behavior[61] and openness to new experi-ences.[62] High levels of creative achievement have been shown to be associated with the exploratory-excitability component of novelty seeking.[63] Curiosity and information seeking also activate the dopaminergic system.[64] For instance, reading high curiosity-eliciting trivia questions was associated with stronger activity in several dopamine-relevant regions including the substantial nigra, ventral tegmental area, and the striatum.[65]

The drive to explore derives from fundamental emotional systems that are similar across species. Drawing from his work on emotions in animals, Jaak Panksepp proposed a model of human personality that ties six primary affective tendencies—playfulness, seeking, caring, fear, anger, and sadness—with

distinct emotional systems. While playfulness is associated
with extraversion, seeking is associated with openness.[66] The
relevance of the dopamine-dependent "motivational seeking
system," or seeking-related affective tendencies, to a disposition that favors creativity is unmistakable, given its characterization as "feeling curious, feeling like exploring, striving for
solutions to problems and puzzles, positively anticipating new
experiences, and a sense of being able to accomplish almost
anything."[67]

Interestingly, openness to new experience and extraversion
are the two dominant personality traits that are associated
with the dopaminergic system. Extraversion is associated with
the seeking of gratification from outside the individual and
reflects the disposition of being outgoing, sociable, openly
expressive, gregarious, assertive, and enthusiastic. The link
between dopamine and extraversion is thereby tied to reward
sensitivity and associated positive emotions.[68] Openness to
experience, in contrast, reflects the disposition to be intellectually curious, imaginative, perceptive, attentive to inner
experiences, aesthetically sensitive, adventurous, and challenging of the status quo. The association between dopamine
and openness is drive-related, as it draws on novelty-seeking
and cognitive exploration.[69] With regard to their joint role
in facilitating psychological function, extraversion is held
to influence an individual's motivation to perform challenging cognitive tasks, while openness plays the key role in the
actual task performance.[70] The involvement of dopamine in
both extraversion and openness has been taken as the basis for
proposing a "metatrait" comprised of both these traits. This
metatrait is referred to as "plasticity" or "engagement" and is
"hypothesized to relate to an individual's basic need to incor-
porate novel information from the environment."[71]

The exploration theory of dopamine proposed by Colin DeYoung focuses in large part on the key involvement of the neurotransmitter in the specific detection of only uncertain and unpredictable motivationally relevant events (as opposed to all events). This uncertainty is held to have inherent incentive reward value that drives exploration-based cognition and behavior. According to this theory, although humans are generally motivated to reduce uncertainty in their daily lives, particularly when faced with the likelihood of aversive or undesirable outcomes, something different occurs when faced with uncertainty that is experienced as interesting: "Exploration is used not only to transform the unknown into the known, but also the known into the unknown."[72] Value-coding and salience-coding dopaminergic neurons are differentiated from one another here. "The value system seems likely to drive unprompted, but potentially fruitful, behavioral exploration of the social and physical world, whereas the salience system seems likely to drive spontaneous innovation and cognitive exploration."[73] While there is very little empirical work that directly assesses exploratory behaviors in the context of dopamine and creativity, the limited evidence on hand paints a mixed picture at best. For instance, individuals with high propensity for openness to experience were found to actually perform better on divergent thinking tasks following the ingestion of a dopamine blocker.[74] So the need to consider added variables, including the baseline level of exploration traits, arises here too.

This emphasis on the salience system in the exploration theory draws upon another key idea about the overarching function of dopamine neurons, namely that they signal "salience"—what information is relevant and needs to be the focus of attention and action.[75] This idea is derived from evidence that shows that dopamine (DA) neurons "are activated under conditions of salient environmental change, conditions that require an organism to (i) become responsive to environmental stimuli, (ii) prepare for the possible output of high levels of behavioral activity, (iii) maintain a working memory representation of the just-encountered event (i.e. the event that caused DA[dopamine] neurons to release large quantities of DA into terminal regions)."[76]

These are situations in which salience is "exogenously" elicited via changes in the external environment of a person. But salience can also be "endogenously" elicited through changes in the internal milieu of a person, such as when transformations occur within the conceptual realm of a person's thought processes. This is most obviously experienced when one has an unexpected insight while thinking about something. The phenomenology of an insight, eureka, or "aha" moment is such that it is characterized by a sudden recognition of a novel solution, understanding, or meaning, despite no real change in the external milieu. The mind is alerted to the new meaningful information because of its sudden apparent salience that comes about by some form of conceptual restructuring. The experience of insight during creative problem-solving has in fact been shown to elicit activity in a wide set of brain regions, including dopamine-relevant areas like the striatum and the ventral tegmental area.[77]

Concluding Thoughts

Most of the evidence on hand that relates dopamine function to creativity is indirect and has given rise to a rather complex picture, in which the precise nature of the relationship has proved resistant to unraveling. This is also true of more direct physiology-based examinations of this relationship.

For instance, the first study to investigate the genetics of creativity demonstrated a positive correlation between genetic polymorphisms in dopaminergic systems (dopamine receptor D2 gene: DRD2) and serotonergic systems (tryptophan hydroxylase gene: TPH) and divergent thinking.[78] However, other studies since then have reported mixed findings. Differences occur as a function of individual differences in gender, personality variables, and intellectual factors,[79] as well as creativity output measures and the type of dopamine genes alongside their interactions.[80] For instance, the 7R polymorphism in the dopamine receptor D4 (DRD4) gene is associated with conflicting findings in relation to novelty seeking[81] as well as poor divergent thinking.[82] Ambiguous findings therefore appear to be the rule in examinations of the relation between dopamine genes and creativity-relevant markers.

Some studies have used structural neuroimaging to examine focal anatomical differences, such as the size of brain regions, which can be correlated to behavioral metrics to make a case for the association of a specific region to a psychological function.[83] Several have shown a positive correlation between creative performance on divergent thinking measures and the volume or tissue density of core subcortical regions of the dopaminergic system including the ventral tegmental area and the substantia nigra,[84] thereby implicating the importance of both the nigrostriatal and the mesocorticolimbic pathways. However, studies using larger samples of participants have shown that such findings are not always reliably replicated.[85] Similarly, reduced dopamine receptor binding in the thalamus has been shown to be associated with higher divergent thinking scores. As the thalamus is an information relay station, one that acts as a gate in filtering and directing information to appropriate target brain areas, reducing the gating threshold of the thalamus would result in higher information flow—which would in turn aid creativity as access to more information would allow more conceptual combinations, and thereby more unique combinations, to surface.[86] However, other studies that have focused on the role of the thalamus in facilitating creative thinking by examining variances in gross structural anatomy have delivered mixed findings.[87] Direct examinations of the link between creativity and dopamine-relevant brain structures have, therefore, also proved inconclusive thus far.

So, what can we conclude about the dopamine-creativity link given this state of affairs?

The one thing that is clear is that we miss the big picture about the nature of the link between the dopamine system and creativity when we approach the question by merely considering linear structural and functional brain correlates, or when we generalize from any one of dopamine's many well-established disparate functional roles in goal-directed behavior and action. A more subtle overarching role for dopamine in creativity is necessary, one that flexibly incorporates all these notions within its folds. After all, the dopamine system at its core shapes motivational control through coding for motivational value and motivational salience, and each of these functions are tied in myriad ways with creativity.[88] Salience indicates what information or ideas are meaningful, relevant, significant, and worth taking note of, regardless of whether they are brought to our attention when taking in the external environment or are evoked as byproducts of our internal mentation. Value indicates the worthiness of the information or ideas and determines whether or not they are worth expending the energy and drive required to approach, explore, and engage with them further. These elements lie at the heart of creativity—there can be no generation of ideas that are novel and satisfying without the ability to perceive meaningful connections (salience coding) and infer their worthiness (value coding).

The evocative words of Marc D. Lewis speak to these considerations of how dopamine may influence creativity: "I think that striatal dopamine embodies the leading edge of much of our intentional cognitive activity—perhaps all of it. After concepts are acquired, they are then called up and used—revised, elaborated, and differentiated into mental constellations of great complexity, power, beauty, and sometimes ugliness. The emotional pulse that propels us through these creative cognitive acts may be so subtle as to barely register in consciousness or to be entirely inaccessible to appraisal and evaluation . . . adult conceptual advances may be motivated by the wisp of reward—a flavor or hue that imbues objects, events, ideas, actions, and even people with a modicum of value. And perhaps without that wisp, that hue, there would simply be no point to engaging with and understanding the world and its components."[89]

My friend and I may be similarly cued by our dopaminergic systems at dawn—perceiving salience and value in the aroma of freshly poured coffee, which will stimulate us to feeling fully awake. But we may be dissimilarly cued by the sound of coffee being poured. Where I merely hear a liquid being poured, my friend hears musical rhythms. And that triggers his creative idea generation. While the wider individual and environmental contexts determine whether perception of salience and value within any given experience is likely to be universal, limited to a sole individual, or somewhere in between, the role of dopamine in it all stays constant. Dopamine facilitates our experience of the unmundane.

7   The Default Mode Network and Creativity (A Myth in the Making)

The various aspects of imagination . . . all involve a common feature: the activity of the brain turned inward to mentally generated representations decoupled from the immediate external environment. That conception of imagination corresponds with descriptions of the brain's default mode network (DMN)—a neural system that has been aptly characterized as the "imagination network."
—Joseph Carroll[1]

The problem with such mindless neuroscience is not neuroscience itself. The field is one of the great intellectual achievements of modern science. Its instruments are remarkable. The goal of brain imaging is enormously important and fascinating: to bridge the explanatory gap between the intangible mind and the corporeal brain. But that relationship is extremely complex and incompletely understood. Therefore, it is vulnerable to being oversold by the media, some overzealous scientists, and neuroentrepreneurs who tout facile conclusions that reach far beyond what the current evidence warrants—fits of "premature extrapolation."
—Sally Satel and Scott O. Lilienfeld[2]

If you do a group average matrix . . . functional connectivity . . . whatever it is . . . seed map or info map . . . you get to a nice correspondence which is like 20 each [per group]. So, fMRI is awesome. It works. It reproduces. The big disconnect is that when you try to relate it to these phenotypes, and you look at the matrices

> comparing the brain data to the behavioral data. . . . They don't
> replicate all that good at a 1,000 [subjects] . . . that is the hard
> thing to wrap your head around. Why is the correlation so differ-
> ent when the imaging data is so good. . . . I don't blame imaging. I
> blame correlations . . . imaging is not the bad guy here.
>
> —Nico Dosenbach[3]

A groundbreaking paper was published in March 2022 that
stands to rattle the foundations of neuroscientific research.[4]
The title of the paper plainly laid out the central conclusion
of the enormously ambitious study: "Reproducible Brain-Wide
Association Studies Require Thousands of Individuals." The
disciplines of psychology and neuroscience have been reel-
ing for more than a decade now from the widely publicized
"replication crisis"—the failure to corroborate a consequential
proportion of published findings.[5] That this was even true of
many of the most enticing and influential findings in these
disciplines, which tend to be published in highly reputed academic journals and garner heaps of attention from mainstream
media, was a severe blow.[6] A flurry of papers on how to conduct
better science followed in response to this crisis as it became
obvious that several commonplace methodological practices
would have to change to ensure these fields were meeting the
rigor required of any discipline claiming to be a science.[7]

However, nothing has quite prepared the field for the conclusions of the March 2022 paper that uncovered the methodological prerequisites of carrying out one specific type of
neuroimaging approach used to study the brain correlates of
psychological functions: BWAS or brain-wide association studies. BWAS are defined as "studies of the associations between
common inter-individual variability in human brain struc-
ture/function and cognition or psychiatric symptomatology."

So, "BWAS relate population variability in brain features (for example, RSFC [resting-state functional connectivity] between two brain regions (edge)) and behavioural phenotypes (for example, cognitive ability)."[8]

BWAS differ from classical functional neuroimaging appro-aches, where brain activity is generally *more directly tied* to psy-chological function, as the neural activity being examined is brain engagement that is specifically elicited in real time dur-ing the performance of a given task. To take a simple example: one can make inferences about the neural basis of grammar comprehension by directly comparing brain activity that is elicited when participants make judgments on the correctness of syntactically unsound sentences (e.g., "The shirt was on ironed.") in contrast to syntactically sound sentences (e.g., "The shirt was ironed.").

In contrast, in BWAS correlations are typically carried out between behavioral indices that are derived from performance on a given task (or even responses on a questionnaire) and brain metrics that are *not directly tied* to the performance of that very task (like brain activity over a rest period or the grey matter volume of a specific brain area).[9] An example here would be to making inferences about the brain basis of gram-mar understanding by correlating behavioral performance on a grammar test with resting-state brain activity recorded from a language-relevant brain area.

What the paper essentially illustrated was that neuroim-aging studies that followed the BWAS approach, and thereby derived conclusions about brain-mind functions by carrying out correlations between behavioral indices and brain metrics, generally required a minimum of two thousand participants for those conclusions to be deemed reliable. This is a severe problem because the typical sample sizes seen in brain-wide association studies are nowhere near that number. In the field of creativity in particular, studies that examine correlations between performance on widely used creativity-relevant measures and structural brain indices (such as white matter or grey matter volume) or functional brain indices (like resting-state or task-related activity or connectivity) sample sizes in the largest studies are well below two hundred. So even in the best-case scenario, the sample sizes are less than 10 percent of what appears necessary for deriving reliable insights.

This is a matter for real concern in the context of the myths/truths of creativity because the current dominant model in contemporary research that is advocated as explaining the brain basis of creative thinking is in large part based on BWAS research. This model places a special emphasis on the role played by the default mode network (DMN) in orchestrating imaginative thought processes and novel idea generation in creativity. In doing so, the model has captured the attention of the scientifically engaged wider general public, and this view of the central importance of the default mode network in creativity is now taken to be fact.[10] This is despite the lack of consistent evidence for the involvement of the default mode network in creativity from studies that have adopted non-BWAS approaches. This final chapter therefore explores what I characterize a myth/truth in the making—the case for the default mode network's preponderant role in creative ideation. And we begin with where it all started: the discovery of the default mode network.

The Default Mode Revolution

In the mid-1990s, a series of astute observations were articulated during the groundbreaking nascent phase of concerted academic neuroimaging research in relation to what was then seen as a puzzling phenomenon. A basic tenet of functional neuroimaging research is that increased brain activity in a select brain region during the performance of a given task is evidence for the role of that brain region in the mental processes that are involved when carrying out that task.[11] Early neuroimaging scientists therefore focused almost entirely on the increases in brain activity with reference to a task of interest, often in comparison to an undemanding baseline, like passive viewing conditions (e.g., staring at a fixation cross on a screen) or rest. The observations that led to the discovery of the default mode network resulted from paying attention to and discerning meaningful patterns from findings that were typically disregarded by most.

First came the observation of positive correlations between task-elicited brain activity in regions of the motor cortex that control hand movements, and baseline brain activity (low-frequency temporal fluctuations) in the exact same regions (as well as others involved in motor control) when movement was not being elicited—that is, while remaining motionless during rest. This gave rise to the proposal that such "correlation of low frequency fluctuations, which may arise from fluctuations in blood oxygenation or flow, is a manifestation of functional connectivity of the brain."[12]

Around the same period, the first meta-analyses of brain engagement on attentionally demanding tasks were published. These showed not only the expected "task-related increases,"[13] but also unexpected "task-related decreases"[14] in brain activity within a distributed set of regions. This latter pattern was particularly intriguing as the same set of brain regions kept surfacing in displaying this signature pattern of task-related decreases across highly dissimilar tasks. In the words of Marcus

Raichle who coined the phrase "default mode": "This consistency with which certain areas of the brain participate in these decreases made us wonder whether there might be an organized mode of brain function that is present as a baseline or default state and is suspended during specific goal-directed behaviors."[15] And the findings from his lab confirmed these predictions. It was shown, for instance, that while one facet of brain metabolic activity as measured by oxygen extraction fraction (OEF) was homogeneous across the brain during rest, another facet of brain metabolic activity as measured by cerebral blood flow (CBF) showed striking heterogeneity across different regions of the brain, such that there was greater blood flow to regions of the brain that are typically associated with "task-related decreases."

Then came the first evidence of functional connectivity in resting-state brain activity between the default state regions.[16] The region of focus was the posterior cingulate cortex (PCC), which showed the highest task-related decreases in brain activity during the performance of a working memory task. The PCC is a brain region that is located on the medial parietal wall of the brain (which used to be referred to as the "medial mystery parietal area" by early default mode researchers who did not quite know what to make of its activity patterns).[17] Low-frequency resting baseline activity (elicited when participants were instructed to close their eyes and think of nothing in particular for a period of four minutes) in the PCC was functionally coupled to the same type of activity in several of the now established core regions of the default mode network. These are now known to encompass the ventral and dorsal medial prefrontal cortex, the inferior parietal lobule, the lateral temporal cortex, the posterior cingulate and retrosplenial cortices, and the hippocampal formation.[18]

From the get-go it was apparent that the type of task mattered. It turned out that default mode regions did not always show decreases as a response to every type of cognitive task. In fact, these regions actually showed task-related *increases* when the tasks to be performed involved engaging in conceptual processing, such as in those necessitating semantic knowledge retrieval[19] and self-referential mentation.[20] This fit with evidence showing that these regions were involved when engaged in task-independent and undirected thoughts as well, that is when peoples' minds presumably wandered and drifted to spaces and issues of interest to themselves.[21] In fact, in one of the earliest studies to adopt the BWAS in this field of study, mind-wandering frequency was found to be significantly correlated with activity in default mode areas.[22] These ideas were further affirmed by studies that carried out retrospective experience-sampling procedures, where participants are asked after the fact what they were thinking about during rest periods while their brains were being scanned. One study found that participants only reported spending 10 percent of the time reflecting on the stimuli they had just seen and only 5 percent of the time did they report having a blank mind. They mostly reflected on the past and the future, and the frequency of such reflections was found to be correlated with functional connectivity of brain regions within the default mode network.[23] Such findings led to the conceptualization of the default mode network as an internal mentation network and the seat of spontaneous cognition.[24] The brain activity patterns associated with the state of "rest" thereafter became the subject of concerted inquiry in relation to a vast array of variables, including creativity. What's more, this discovery that intrinsic connectivity could be detected at rest for diverse brain networks (not just the default mode) between the implicated brain regions in each network[25] led to a rapid surge in the use of resting-state-based analysis in neuroscientific research.

Early Default Mode Ventures in Creativity

The first neuroscientific paper to overtly put forward a default mode explanation in relation to creativity was published in 2011. Sixty-three people were tested in two separate testing sessions: a behavioral session where their performance on a creativity test was assessed, and a functional neuroimaging session where their brain activity was recorded while they performed an unrelated (noncreative) working memory task. Correlations were carried out between the brain activity elicited during the neuroimaging session and performance on the creativity measure assessed during the behavioral session. A negative correlation was found, such that high creative performance was associated with lower brain activity in the precuneus, a parietal region in the posterior medial wall of the brain. The researchers identified this to be a region of the default mode network, which then informed their interpretation of the finding: "The magnitude of TID [task induced deactivation] in the default mode network is considered to reflect the reallocation of cognitive resources from networks irrelevant to the performance of the task. Thus, our findings may indicate that individual creativity, as measured by the divergent thinking test, is related to the inefficient reallocation of attention, congruent with the idea that diffuse attention is associated with individual creativity."[26] This pattern of choosing a network-level interpretation of brain activity over a region-level interpretation, despite evidence of the involvement of only one brain region from the many that form a given brain network, is a curious but common practice in neu-roimaging literature. And the glaring subjectivity inherent to these interpretations becomes even more apparent when one notices that the network-level explanation provided doesn't quite correspond with the functions accorded to that brain structure when following a region-level explanation.[27]

That same year Nancy Andreasen published a perspective article in which she highlighted how her work from the mid-1990s was prescient in recognizing that the brain is never inactive and is perpetually engaged in "random silent episodic thought" (REST) processes, even during periods of actual rest. She was therefore very likely to have been the first to attribute processes of undirected thought and free association to the default mode network. She postulated that "the brain regions most likely to be involved in the creative process are the association cortices, those brain regions that are most active during REST when a person is engaged in free-ranging and uncensored thought."[28]

In 2012 came the first study that made a case for combined influence of the default mode network and the executive func-tion network in creativity by examining brain activity patterns that were directly elicited during a creative task. Fifteen art and design students were given the task of creating six book cover designs. For each of the six trials, they were presented with a summary book description for forty-five seconds after which they were given thirty seconds to draw or write their ideas for a cover for that book. This "generation" phase was followed by a brief break involving the tracing of lines, and then came the "evaluation" phase in which the students had twenty sec-onds to draw or write their appraisal of the book cover idea they had generated. This cycle was repeated five times for each book cover trial at the end of which participants were asked to rate their own success at being able to generate and evaluate their own output. The authors carried out a direct comparison of the brain regions that were selectively involved in each of these phases. The generation phase saw the involvement of several structures including the hippocampus. Although the hippocampus is a core region of the default mode network, the authors interpreted the activation seen in the generation phase as one that reflects medial temporal lobe (MTL) memory functions, presumably because of the co-activation of other MTL relevant regions, such as the parahippocampal formation, during the same phase. In contrast, the evaluation phase saw the involvement of several brain regions, many of which belong to the default mode and executive control networks. The authors proposed "that the medial temporal lobe may be central to the generation of novel ideas and creative evaluation may extend beyond deliberate analytical processes supported by executive brain regions to include more spontaneous affective and visceroceptive evaluative processes supported by default and limbic regions." They thereby concluded that creative thinking "recruits a unique pattern of opposing neural and cognitive processes."[29]

A similar conclusion was reached in an earlier 2008 study that examined the neural basis of improvisation in six professional musicians. However, this conclusion was derived from almost diametrically opposing findings. In comparison to the production of overlearned musical sequences, musical improvisation was associated with deactivation across specific core regions of the default mode network (dorsomedial prefrontal cortex), the executive control network (dorsolateral prefrontal cortex), and limbic system (anterior insula), alongside activation of the anterior-most regions of the medial prefrontal cortex (default mode network). The authors summarized the findings thus: "This unique pattern may offer insights into cognitive dissociations that may be intrinsic to the creative process: the innovative, internally motivated production of novel material (at once rule based and highly structured) that can apparently occur outside of conscious awareness and beyond volitional control." This is rooted in the idea that the lateral prefrontal cortex which ordinarily "support[s] self-monitoring and focused attention" processes when deactivated "may be associated with defocused, free-floating attention that permits spontaneous unplanned associations, and sudden insights or realizations."[30]

I have delved into the specifics of these four papers not only because they are among the most influential—or were the first of their kind—in the field of creative neurocognition, or featured noteworthy shortcomings. (The latter is often an unavoidable feature of explorative and daring investigations within a new field of study.) It is because they are illustrative of the problem of subjective interpretations that are inescapable in neuroscience but can be compounded exponentially when there are no checks in place for confirmation and overgeneralization biases. For instance, each of the four papers emphasizes different roots—from specific brain regions to whole brain networks—as underlying associative thinking in creative idea generation.[31] Any study that followed could therefore retrofit the interpretations of their findings as being in alignment with the same hypothesis regardless of the specifics of the results. Moreover, owing to the ambiguity associated with the brain region descriptors and their functional roles within the papers themselves,[32] these studies are often incorrectly cited.[33] These (alongside others noted later) are some of the ingredients that propagate the cycle and spread of misattributed and misgeneralized findings in the study of creative neurocognition.

The Default Mode Echo Chamber

The steep rise of default mode explanations for creativity after this period coincided with the publication of three theoretical papers that made a case for the central role played by the default mode network in creative ideation. It commenced with Rex Jung's 2013 model, where this network was deemed responsible for blind variation mechanisms that allow for novelty during idea generation. In contrast, the central executive network was held to regulate the selective retention mechanisms that allow for the refinement and evaluation of the most viable ideas.[34] That creativity is orchestrated by both spontaneous and controlled processes of the brain, with the default mode involvement in the former process, was emphasized by Leh Woon Mok.[35]

Roger Beaty and colleagues integrated both these theoretical ideas in characterizing the default mode network as a spontaneous generative system while the central executive network was dubbed the deliberate control system. An important extension was their emphasis on the dynamic interplay between the two networks during creative ideation, which is also an idea that echoes earlier theoretical proposals in the field of creativity.[36] However, they put forward two conflicting views within the very same paper. They noted, on one hand, that the two networks on occasion work in tandem or are co-activated during *both* phases of creative ideation, such that "the control network may couple with the default network during idea generation or evaluation to constrain cognition to meet specific task goals." Yet in another part of the paper, they emphasize a view that parallels Jung's idea, that each network subserves different phases: "We suggest that the default network contributes to the generation of candidate ideas (potentially useful information derived from long-term memory) in light of its role in self-generated cognition (e.g., episodic memory). Yet, the control network is often required to evaluate the efficacy of candidate ideas and modify them to meet the constraints of task-specific goals."[37] So, here again, studies finding evidence for either single or dual network involvement in each phase of creative ideation can be claimed as showing support for this theoretical model.

A slew of functional neuroimaging studies using a variety of complex data analytic methods have now been published that indicate the involvement of both these large-scale brain networks in creative thinking, with the default mode network being generally ascribed the generative or associative functional role in ideation process. A dire problem for this claim of a notable generative role for the default mode network, though, is that the evidence put forward in support of this assertion is largely based on brain-wide association studies (BWAS), or correlations between behavioral indices and brain metrics.[38] Remember that this is the kind of neuroimaging analysis that needs a sample size of around two thousand to be reliable. The general sample sizes seen in the field of creative neurocognition (and indeed most other fields of neuroscience as well) are far smaller. Among the BWAS cited in this chapter, for instance, sample sizes ranged from thirteen to 186.

This raises serious questions about how much reliable support is actually there for the network dynamics (specifically between the default mode and central executive networks) model of creativity. This is particularly so when considering that non-BWAS approaches do not often show evidence for strong default mode network involvement in creativity. In fact, meta-analyses of findings from such studies consistently implicate the importance of the semantic cognition network,[39] and not the default mode network, which is in line with both classical and contemporary theories of the creative mind.[40] In fact, default mode network operations like episodic thinking are not associated with the defining attribute of creativity: "originality" during idea generation.[41] The same is true of other default mode network processes, like mind-wandering, which appear to be uncorrelated with creative performance, behavior, or achievement.[42]

Moreover, while much is made of the role of the default mode network in generation versus evaluation stages of the creative ideation process, and the coupling of this network activity to that of the central-executive-network, it should be noted that most studies in the field of creative neurocognition are not designed to evaluate temporal aspects of these network interactions.[43] This is because (1) they customarily employ divergent thinking tasks in which these phases are not separated to assess creativity,[44] and (2) functional neuroimaging data do not typically possess the requisite temporal resolution to allow for fine-grained temporal analyses, particularly in the context of complex processes like creative cognition.[45]

There are two further crucial points to note here that cast a shadow on much of how research trends progress in the field of creative neurocognition, notwithstanding the newly recognized problems of adopting the BWAS approach.[46] One is that neuroscientists regularly reach for network-level explanations, even when their findings do not show the involvement of all (or even most of) the regions of that network they are advocating for. The second is that brain regions that are the focus of interest in creativity studies generally do not belong to any given single brain network, but actually constitute core regions in multiple brain networks. For instance, the posterior cingulate cortex is part of the default mode network and the semantic cognition network,[47] whereas the dorsal anterior cingulate cortex belongs to both the central executive network and the salience network.[48]

These practices open the door for radical interpretative subjectivity. Let's take the case of the clever book cover design study that was detailed in an earlier section of this chapter. Even though the authors themselves interpreted the hippocampal activation as reflecting the involvement of the medial temporal lobe memory system during idea generation, this paper is cited by others as providing evidence for default mode activity in creative idea generation *because* the hippocampus is also a default mode network region.[49] So hippocampal activation during creative ideation can be attributed to the medial temporal lobe memory network by one set of researchers and the default mode network by another group. Using the same tactic, one can just as easily make a case for the involvement of other brain networks by focusing on any one of the other co-activated brain regions (e.g., the activation of the inferior frontal gyrus could reflect semantic network involvement).[50] Couple this interpretative subjectivity and the practice of not discussing competing explanations with the fact that network-level explanations are forwarded even when the findings often only implicate single brain regions. What results is a severe echo chamber problem where anything goes as long as it falls in line with the dominant ideological narrative. Academic groupthink thereby not only becomes pervasive; over time it becomes established as the norm.

Default Mode Dilemmas and Its Unspecial Role in Creativity

Resting-state-based analyses are the dominant neuroimaging analysis approach employed when making a case for the critical role played by the default mode network in creativity. The incentives to employ resting-state analyses to examine default mode activity are substantial, as the time taken to record such data is in the order of a few minutes. So, it is a simple, quick, and cost-effective method to obtain a functional MRI dataset: "The appeal of the technique lies in its simplicity. A brief MRI data set acquired in resting subjects is sufficient to explore diverse brain systems."[51] What's more, not only can the resting-state MRI data itself be analyzed multiple times using a range of complex analytical tools, but because different cognitive or behavioral measures are also included as part of the assessment (either before or after the imaging session), BWAS analyses can be carried out in relation to each of those measures by correlating them with the resting-state brain activity. This makes for a fodder dataset for multiple publications.

Several issues crop up, however, with the adoption of resting-state network (RSN) analyses, particularly in comparison to the classic alternative of cognitive function network (CFN) analyses. This is when correspondence between RSN and CFN findings in some specific domains of study[52] is taken as adequate grounds to assume generalizations across all domains of study. The assumption then is that RSN can be used to substitute CFN in all domains. Not only is there "now abundant, direct evidence that patterns of co activation, or connectivity, of RSNs are significantly and meaningfully different than those of the CFNs recruited by a variety of cognitive tasks"—creativity being one such example—"a central problem with the use of RSNs in cognitive neuroscience is that they do not shed light on the neural instantiation of cognitive processes."[53]

Notwithstanding this and other troubling conceptual methodological issues associated with the use of resting-state analyses,[54] there are also technical problems to come to terms with when using these techniques. It has been long known that functional connectivity MRI measures are highly sensitive to physiological confounds like respiration rate, head motion, blood pressure, and cardiac rhythm, which give rise to technical artifacts in the data.[55] Good disclosure practices in how such physiological confounds have been dealt with are far from the norm.[56] Moreover, even the type of instruction given as to what participants are supposed to think about (or rather not think about) during rest directly impacts resting-state brain activity.[57] Given that ongoing internal mentation influences resting-state brain activity, recent guidelines emphasize that participants need to be asked afterward what they were thinking about while their brains were being scanned as they were resting.[58] It is remarkable that posttesting introspective reporting techniques have not been employed so far to corroborate the generativity hypothesis for the default mode network in relation to creative ideation. A recent critique of resting-state analyses offered the following cautious guidance: "While rest likely still has a place, the choice to collect resting-state data should be deliberate in the same way that the choice of any other task is deliberate—driven by where it can add true value, and not just be an easy default."[59]

What is often emphasized when referring to brain network dynamics in creativity is the special conjoined involvement of the default mode and central executive networks: the atypical coupling of brain networks that are typically anti-correlated.[60] However, such coupled brain network activity patterns are not limited to creativity, and indeed occur when performing tasks that assess goal-directed thought, such as working memory and autobiographical planning,[61] as well as during tasks that involve their diametric opposite, such as undirected thought or mind-wandering.[62] For instance, the default mode network (DMN) is even active during high demand working memory tasks, where it is believed to encode task-relevant information: "We propose that the DMN may be recruited whenever large changes of cognitive context are required. This may apply in complex cases of self-referential processing, mind wandering etc., but also in relatively simple acts of cognitive or executive control. The DMN, widely seen as a 'task-negative' network, may respond positively to any task which demands a switch from one broad context to another."[63] Change in context here refers to change in informational meaning rather than a change in informational form. The default mode network is sensitive to the changes of context but not changes of informational form. Interestingly, this links back to an early alternative hypothesis about the default mode function, namely that it is also associated with "stimulus-oriented thought" and not just "stimulus-independent thought," depending on the task circumstances.[64] As it stands, the customary characterization of the default mode network as a purely internal mentation or internal-oriented cognition network[65] is increasingly being called into question.[66]

The mounting evidence showing that the default mode network is not only involved in internal mentation but also selectively responsive to meaningful changes in the external environment has given rise to a novel hypothesis about its general function, namely that it is "dynamically integrative." Our knowledge, experiences, and beliefs about ourselves and the world at large—the source material for our internal mentation—influence the way we interpret our ever-unfolding external environment. And, conversely, the happenings of our external world shape and inform the way we think, feel, and believe. The suggestion here is "that the DMN is 'default' because it is central for integrating external and internal information, allowing for shared communication and alignment tools, shared meanings, shared narratives and, above all, shared communities and social networks. This is what people continuously and naturally do by default."[67]

So not only is default mode network activity neither specific nor special to creativity, even the "generativity hypothesis" in relation to the default mode network is called into question, given the evidence of this network's role in making sense of the world and relating it to what is already known. This means that the default mode network tracks, confirms, and updates expectations. A system that cues for novel generativity, in contrast, would necessarily be one that breaks expectations and expands into unknown conceptual spaces.

A similar expectation-confirming role for the default mode network was proposed a few years prior in a neurophilosophical model of the human imagination. A central role was accorded to the default mode network in underlying "intentionality-based forms" of imagination, a category so named as it is activated by "processing that is predominantly recollective in nature with a view to establishing the best possible explanation of a situation or event in question. This is brought about by means of spontaneous access to an extensive and diverse repertoire of relevant knowledge when processing such contexts. The best or most plausible explanation is the one that fits best with what is already known in terms of oneself and/or one's world-view."[68] In contrast, generating an original response during creative ideation is an example of "novel-combinatorial forms" of the imagination. Such forms necessarily draw on the additional activity of non-default mode networks, such as the semantic cognition and central executive networks, as they necessitate resisting the path of least resistance by either overriding the prepotent cycles of conceptual reasoning or considering previously ignored perspectives.

Concluding Thoughts

The parallels between the unfolding of the first myth/truth explored in this book (the creative right brain) and the current myth/truth in the making are rather striking. Despite the involvement of both hemispheres in creativity (truth), the emphatic focus on just the right brain (which perhaps was an understandable tactic as it was the half that had been theretofore disregarded and even painted as deviant) led over time to the establishment of the now unshakeable idea that creativity resides in the right brain (myth). Similarly, in emphasizing the role of the default mode network over and above other brain networks in the generation of creative ideas, despite the clear evidence to the contrary, one sees how a skewing of the narrative leads to the establishment of a simple story.

Unfortunately, no phenomenon as complex as the human capacity for generating novel and satisfying ideas can be distilled down to a simple narrative. To understand the reality for what it is, it is necessary that we override our overwhelming predisposition to see and explain the world in dualities.

While dual models abound in psychology generally and are not specific to the field of creativity,[69] it is critical to both acknowledge and bear in mind that dualistic proposals—that the divergent/generative/open-mode/right-brain/associative system is in opposition to or more crucial for creativity than the convergent/evaluative/close-mode/left-brain/evaluative system —are necessarily extreme simplifications. Their explanatory power lies in conveying big picture take-home messages, not mechanism-level understandings. To achieve the latter is a formidable challenge but is it one that can only transpire by acknowledging the enormity of the quest and doing what is necessary to keep resolutely on the path of real discovery, away from the neurofashions of the time. Although several researchers have detailed conceptual and methodological problems that are endemic to the field of creative neurocognition and some have provided guidelines on tackling such issues,[70] these alerts go largely unheeded, and research continues as though the alarm bells were not sounded.

Interestingly, the idea that associative/generative and executive/evaluative processes underlie creative thinking can be traced back to the psychoanalytic tradition.[71] In what was perhaps the earliest formal dualistic model of the creative brain, Sigmund Freud distinguished between "primary process" and "secondary process" thinking. Primary process thinking was marked by a mode that was fantasy-oriented, free associative, concrete, and analogical whereas secondary process mode was logical, goal-oriented, abstract, and reality-based. Novelty in the ideation process was ascribed to having atypical access to primary process thought while secondary process thought allowed for the shaping of these ideas into a form that would be deemed meaningful and palatable to others.[72] But Freud was prescient in recognizing the danger of reaching hasty conclusions that could easily occur when generalizing from simple dualities. The warning he sounded makes for a fitting end to this last myth/truth chapter: "[The artist] is not the only one who has a life of phantasy; the intermediate world of phantasy is sanctioned by general human consent, and every hungry soul looks to it for comfort and consolation. But to those who are not artists the gratification that can be drawn from the springs of phantasy is very limited; their inexorable repressions prevent the enjoyment of all but the meager day-dreams which can become conscious. A true artist has more at his disposal. First of all he understands how to elaborate his day-dreams, so that they lose that personal note which grates upon strange ears and become enjoyable to others. . . . Further, he possesses the mysterious ability to mold his particular material until it expresses the ideas of his phantasy faithfully."[73]

Afterword

> [A] concept that is used deliberately, not blindly, in the science
> for which it was created, where it originated, developed, and was
> carried to its ultimate expression, *is blind*, leads nowhere, when
> transposed to another science.
>
> —Lev Vygotsky (1896–1934)[1]

Here is a well-known fable about research. A scientist is testing
how far a frog can leap upon instruction. He commands the
frog to leap. The frog dutifully complies. The scientist measures the space traversed and makes a note—"fully able frog, 8
inches jump." The scientist then amputates one leg of the frog
and commands the frog to leap once again. The frog responds,
and the scientist notes the distance covered—"one leg gone: 5
inches jump." The scientist proceeds to amputate yet another
leg of the frog, and the cycle of events repeats until the last
leg is removed. When the scientist now commands the frog to
leap, the frog makes no movement. He repeats the command,
yet the frog remains still. He notes in his observation log—
"four legs gone: the animal is deaf."

This tale illustrates the folly of scientific conclusions that
result from (knowingly or unknowingly) adopting tunnel

vision in the pursuit of knowledge. Even adhering to a scru-
pulously tidy methodological chain in one's logical analysis
can lead to startlingly erroneous conclusions that, although
obvious in the case of the frog parable, unfortunately often go
undetected in the real world of quantitative and qualitative
research, despite best intentions.

There are many grounds for why this happens. Some fields
of study are potentially more at risk in being blindsided by the
weight of knowledge and individual expectations because the
phenomena under study are utterly enchanting and elusive
to begin with.[2] Creativity is, without doubt, one such field.
Scientific explanations that are not immediately captivating
to the reader perhaps somehow feel unsatisfying to ascribe to
phenomena that are as awe-inspiring and wonder-inducing as
feats of creativity. In some situations, the explanations themselves are so compelling for one field of study that they are
blindly generalized to others, leading to misinformation that
is very difficult to displace once it has set in, as it becomes part
of a larger narrative that has its roots in specific truths.[3]

The intention behind the research, which is fundamental to
any scientific investigation, is perhaps the most critical factor
to consider. What is the question guiding the quest? We do
not look for answers to complex phenomena with zero prior
expectations. We search and research because we want to know
more about something that has stirred our curiosity. That, in
and of itself, guides the direction of our quest and the approach
we choose to adopt. When we discover something that makes
good sense, we feel the pleasure that accompanies the dawn of
a new understanding that aligns with our expectations. When
we dig deeper and find more that fits this initial picture, a narrative develops, and we experience the continued positivity that
comes with getting things right, as well as a steady increase of

confidence in our understanding of the phenomenon. What we gain over time is a great story, one that helps us navigate the fascinating puzzle. A story that goes some way in explaining the persistent link between creativity and mental illness, or the right brain, or psychedelic drugs, and so on.

A good story is more compelling than the truth.[4] Ask any journalist. Human beings are extraordinarily social, and storytelling (and therefore story-listening) is an intricate part of our way of life. We use stories and narratives to communicate information about ourselves and others. Through it we pass on our histories, cultures, values, identities, ideas, and more. In fact, we not only have a penchant for it, we are also hardwired to tell and receive stories as a species.[5] A fitting narrative that has all the elements of a good story, and aligns with what we think to be true, is going to be more readily accepted than something that does not.[6]

We can see an illustration of this principle from a series of papers that explored the influence of neuroscientific evidence in popular explanations of psychological phenomena. By and large, such examinations are guided by the assumption that information couched in brain-based frameworks is enormously appealing and leads to uncritical acceptance of the views, particularly by a naïve general public. A prominent example was a case that was made for the "seductive allure of neuroscientific explanations" from a study demonstrating that explanations were deemed more satisfying by nonexperts when neuroscientific content (which was logically irrelevant) was included in the information, than when it was not.[7] Another study went further and argued for the "selective allure of neuroscientific explanations" as the researchers found that this sway of influence only holds when the information being peddled corresponds with one's own expectations.[8]

Studies like these can garner a lot of media attention, which contributes to the rapid spread of such views. But as more research teams stepped in to explore this question, what followed was the failure to confirm such findings. A damning verdict on the issue was declared in a paper titled "The Seductive Allure of the 'Seductive Allure,'" where it was proposed that it is because the scholars themselves believe that neuroscientific evidence is unduly influential that they are inadvertently inclined toward designing studies and interpreting information in a way that confirms their presuppositions.[9]

What this illustrates is the problem of confirmation biases in scientific reasoning,[10] which is inevitable given that scholars typically aim to find evidence that validate their ideas. When scientists do not seek out refutations of their views, and competing hypotheses are neither considered nor tested, what results is a severe echo chamber problem. This is because the impact of such practices is not only limited to the research being conducted. Research findings are disseminated; these views then spread and become commonplace. This can promote academic groupthink in any field, which in turn compromises (and perhaps is even made worse by) the peer-review process, where gatekeepers who subscribe to the dominant view can block alternate viewpoints from being aired.[11] Indeed, "status bias" is rampant in the peer review process as author prominence plays a critical role in how easily a paper is given the green light for publication.[12]

These are some of the many factors that have contributed to the crisis of reproducibility and replication of research findings not only in the life and social sciences, but also in the natural sciences.[13] And they explain, at least in part, how myths about the workings of our minds come to be established.

Each of the seven chapters in this book was structured to uncover the kernels of truth underlying seven myths about the creative brain, as well as the chain of events that led to specific narratives being propagated. In adopting this approach, my direct aim was to inform readers about the bigger and— dare I say—more exciting stories surrounding these dominant ideas. By reevaluating these stories in this manner, I also tried to serve my oblique aim for this book, which was to build a case for the value in embracing the gray zone of the complex unknown over conveniently simple narratives. I believe this is necessary because we live in an era where mediums of pithy communication are increasingly favored for the dissemination and reception of ideas (academic and otherwise), and easily digestible soundbite science is ceaselessly demanded and promoted by news media and popular podcasts. Unfortunately, these are the very ingredients that encourage the formation and perpetuation of myths.

There are many resources for the research community that provide guidance on how to conduct good science and create a culture that promotes sound practices.[14] Unfortunately, apart from some popular science books that explore the importance of being more skeptical of information being pitched by scientists or even one's own reasoning processes,[15] there are few resources we can refer to for guidance on how to overcome the many mental hurdles in place that cause us to slide toward easy explanations.

Perhaps this is because it is inevitable. One part of coming to terms with this problem, then, is being able to recognize the unobjective predispositions we are endowed with as human beings and to accept our fragility and lack of control. The other part is to recognize the strengths we have in our capacity for "metacognition," which is the awareness we have about our own thoughts, and the understanding of *why* we have those thoughts.[16] So, even if we are physiologically unable to override the vicious cycle of informational persuasion for the most part, we can (and need to) exert considered efforts in our own ways as individuals to take pause whenever possible and ask curiosity-driven questions. What is missing from this story? What is not being explained here? Where does this explanation lead us? The challenge here is that we need to pose such questions even when the story "feels" beautiful, truthful, or satisfying. The question we should dare to ask ourselves even when being dazzled is this:

It may be the poetic truth, but what about the larger truth?[17]

Notes

Introduction

1. Bertrand Russell, *Proposed Roads to Freedom: Socialism, Anarchism, and Syndicalism* (New York: Henry Holt & Co., 1919), 147.

2. Havelock Ellis, *Studies in the Psychology of Sex, Vol. V: Erotic Symbolism; The Mechanism of Detumescence; The Psychic State in Pregnancy* (Philadelphia: F. A. Davis Co., Publishers, 1920).

3. Marie-Hélène Huet, *Monstrous Imagination* (Cambridge, MA: Harvard University Press, 1993).

4. Originally introduced in 1573: Ambroise Paré and Michel Jeanneret, *Des monstres et prodiges* (Paris: Gallimard, 2015).

5. Norbert Schwarz, Eryn Newman, and William Leach, "Making the Truth Stick & the Myths Fade: Lessons from Cognitive Psychology," *Behavioral Science & Policy* 2, no. 1 (2016): 85–95, https://doi.org/10.1353/bsp.2016.0009.

6. Michael Shermer, *Why People Believe Weird Things: Pseudoscience, Superstition, and Other Confusions of Our Time* (New York: Henry Holt and Company, 2002), 53.

7. Mathias Benedek et al., "Creativity Myths: Prevalence and Correlates of Misconceptions on Creativity," *Personality and Individual Differences* 182 (November 2021): 111068, https://doi.org/10.1016/j.paid.2021.111068.

8. Betty Edwards, *Drawing on the Right Side of the Brain: The Definitive, 4th Edition* (New York: TarcherPerigee, 2012); Lucia Capacchione, *Recovery of Your Inner Child* (New York: Simon & Schuster, 1991).

9. Renowned critics include Arne Dietrich, "The Mythconception of the Mad Genius," *Frontiers in Psychology* 5 (February 26, 2014), https://doi.org/10.3389/fpsyg.2014.00079; Judith Schlesinger, "Creative Mythconceptions: A Closer Look at the Evidence for the 'Mad Genius' Hypothesis," *Psychology of Aesthetics, Creativity, and the Arts* 3, no. 2 (2009): 62–72, https://doi.org/10.1037/a0013975; and Robert W. Weisberg, *Creativity: Genius and Other Myths* (New York: W. H. Freeman & Co, 1987).

10. Ullrich K. H. Ecker et al., "The Psychological Drivers of Misinformation Belief and Its Resistance to Correction," *Nature Reviews Psychology* 1, no. 1 (January 2022): 13–29, https://doi.org/10.1038/s44159 -021-00006-y.

11. W. Joseph Campbell, *Getting It Wrong: Debunking the Greatest Myths in American Journalism*, 2nd ed. (Oakland: University of California Press, 2017).

12. An idea that is experienced as being novel and satisfying to the person generating it is an expression of creativity at the level of an individual (or I-creativity). When that experience is shared by others as well, the magnitude of the creative idea increases in alignment with the number of people who undergo the experience, thereby shifting from a purely subjective standpoint to include an increasingly objective frame (mini-c > little-c level > Pro-c > Big-C). At the Big-C end of the continuum are the ideas that are experienced as being novel and satisfying to the larger collective—the kind of ideas that can change the world. These "eminent" ideas are indicative of historical or H-creativity. The foundational sources for these proposed views include the following: Morris I. Stein, "Creativity and Culture," *Journal of Psychology* 36, no. 2 (1953): 311–322, https://doi.org/10.1080 /00223980.1953.9712897; Margaret A. Boden, "Creativity and Biology," in *Creativity and Philosophy*, ed. Berys Nigel Gaut and Matthew Kieran (New York: Routledge, 2018), 173–192, https://doi.org/10.4324 /9781351199797-11; and James C. Kaufman and Ronald A. Beghetto,

"Beyond Big and Little: The Four C Model of Creativity," *Review of General Psychology* 13, no. 1 (2009): 1–12, https://doi.org/10.1037/a0013688.

13. He did so in 1926 in his scholarly work titled "Religion in the Making: Lowell Lectures": "Creativity, n.," in *OED Online* (Oxford University Press), accessed December 8, 2022, https://www.oed.com/view/Entry/44075.

14. Alfred North Whitehead, *Dialogues of Alfred North Whitehead* (Boston: Little, Brown, 1956), 16.

Chapter 1

1. Karl Pearson, *The Life, Letters and Labours of Francis Galton*, vol. 3b (London: Cambridge University Press, 1930), 500.

2. Pearson, *The Life, Letters and Labours of Francis Galton*, 500.

3. Jonathan Evans and Keith Frankish, eds., *In Two Minds: Dual Processes and Beyond*, illustrated ed. (Oxford; New York: Oxford University Press, 2009).

4. A thorough history of the dual brain/mind idea is provided in Anne Harrington, *Medicine, Mind, and the Double Brain: A Study in Nineteenth-Century Thought* (Princeton, NJ: Princeton University Press, 1987).

5. Henry Holland, "On the Brain as a Double Organ," in *Chapters on Mental Physiology* (London: Longman, Brown, Green, and Longmans, 1852), 170–191, https://doi.org/10.1037/12043-008.

6. A. L. (Arthur Ladbroke) Wigan, *A New View of Insanity: The Duality of the Mind Proved by the Structure, Functions, and Diseases of the Brain, and by the Phenomena of Mental Derangement, and Shewn to Be Essential to Moral Responsibility. With an Appendix* (London : Longman, Brown, Green, and Longmans, 1844), https://wellcomecollection.org/works/c6sz3dc2.

7. Paul Broca, "Remarques sur le siége de la faculté du langage articulé, suivies d'une observation d'aphémie (perte de la parole)," *Bulletin de La Société Anatomique* 6 (1861): 333.

8. Carl Wernicke, *Der aphasische Symptomencomplex; eine psychologische Studie auf anatomischer Basis*. (Breslau: Cohn & Weigert, 1874).

9. L. Lichtheim, "On Aphasia," *Brain* 7 (1885): 433–485.

10. J. Brinkman and H. G. J. M. Kuypers, "Splitbrain Monkeys: Cerebral Control of Ipsilateral and Contralateral Arm, Hand, and Finger Movements," *Science* 176, no. 4034 (May 5, 1972): 536–539, https://doi.org/10.1126/science.176.4034.536.

11. J.M.S. Pearce, "Hugo Karl Liepmann and Apraxia," *Clinical Medicine* 9, no. 5 (October 2009): 466–470, https://doi.org/10.7861/clinmedicine.9-5-466.

12. As cited in Stephen D. Christman, "A History of Brain Asymmetry Studies," in *Reference Module in Neuroscience & Biobehavioral Psychology* (Elsevier, 2018), 2–3, https://doi.org/10.1016/B978-0-12-809324-5.21934-2.

13. Harrington, *Medicine, Mind, and the Double Brain*, 83.

14. Harrington, 80.

15. Henry Maudsley, "The Double Brain," *Mind* 14, no. 54 (1889): 161–187.

16. Stephen J. Gotts et al., "Two Distinct Forms of Functional Lateralization in the Human Brain," *Proceedings of the National Academy of Sciences of the United States of America* 110, no. 36 (September 3, 2013): E3435, https://doi.org/10.1073/pnas.1302581110.

17. While the corpus callosum enables inter-hemispheric cortico-cortical communications, other tracts enable within hemisphere or intra-hemispheric cortico-cortical communication, the largest of which include the superior longitudinal fasciculus, the uncinate fasciculus, and the arcuate fasciculus: Norman D. Cook, *The Brain Code: Mechanisms of Information Transfer and the Role of the Corpus Callosum* (London; New York: Methuen, 1986).

18. R. W. Sperry, "The Great Cerebral Commissure," *Scientific American* 210, no. 1 (1964): 42.

19. See Harrington, *Medicine, Mind, and the Double Brain*, 282 and chapter 7, for details on John Hughlings Jackson's (1835–1911) proposal that the left and right hemispheres are both involved in processing propositional thought, with the left brain doing so by means of words (verbal ideas) and the right brain through images (visual ideas and even "tactual ideas"). For more on right hemisphere contributions to language, see Mark Beeman and Christine Chiarello, eds., *Right Hemisphere Language Comprehension: Perspectives from Cognitive Neuroscience* (Mahwah, NJ: L. Erlbaum Associates, 1998).

20. R. W. Sperry, M. S. Gazzaniga, and J. E. Bogen, "Interhemispheric Relationships: The Neocortical Commissures; Syndromes of Hemisphere Disconnection," ed. P. J. Vinken et al. (Amsterdam: North-Holland Publishing Co., 1969), 287, https://resolver.caltech.edu /CaltechAUTHORS:20170414-111911293.

21. Yair Pinto, Edward H. F. de Haan, and Victor A. F. Lamme, "The Split-Brain Phenomenon Revisited: A Single Conscious Agent with Split Perception," *Trends in Cognitive Sciences* 21, no. 11 (November 2017): 835–851, https://doi.org/10.1016/j.tics.2017.09.003.

22. Howard Gardner, "What We Know (And Don't Know) about the Two Halves of the Brain," *Journal of Aesthetic Education* 12, no. 1 (1978): 113–119, https://doi.org/10.2307/3331854.

23. Michael E. Staub, "The Other Side of the Brain: The Politics of Split-Brain Research in the 1970s–1980s," *History of Psychology* 19, no. 4 (November 2016): 259–273, https://doi.org/10.1037/hop0000035.

24. Joseph B. Hellige, *Hemispheric Asymmetry: What's Right and What's Left* (Cambridge, MA: Harvard University Press, 2001).

25. Sebastian Ocklenburg, "Laterality," in *Encyclopedia of Behavioral Neuroscience*, 2nd ed. (Elsevier, 2022), 350–356, https://doi.org/10.1016 /B978-0-12-819641-0.00043-8.

26. Onur Güntürkün and Sebastian Ocklenburg, "Ontogenesis of Lateralization," *Neuron* 94, no. 2 (April 19, 2017): 249–263, https://doi.org /10.1016/j.neuron.2017.02.045.

27. Christine Chiarello et al., "Structural Asymmetry of the Human Cerebral Cortex: Regional and between-Subject Variability of Surface Area, Cortical Thickness, and Local Gyrification," *Neuropsychologia*, The Neural Bases of Hemispheric Specialisation, 93 (December 2016): 365–379, https://doi.org/10.1016/j.neuropsychologia.2016.01.012.

28. Patrizia Bisiacchi and Elisa Cainelli, "Structural and Functional Brain Asymmetries in the Early Phases of Life: A Scoping Review," *Brain Structure & Function*, March 18, 2021, https://doi.org/10.1007/s00429-021-02256-1.

29. Kenneth Heilman, Edward Valenstein, and Robert Watson, "Neglect and Related Disorders," *Seminars in Neurology* 20 (February 2000): 463–470, https://doi.org/10.1055/s-2000-13179.

30. Luca Rinaldi et al., "The Effect of Blindness on Spatial Asymmetries," *Brain Sciences* 10, no. 10 (September 23, 2020), https://doi.org/10.3390/brainsci10100662.

31. Paolo Bartolomeo and Tal Seidel Malkinson, "Hemispheric Lateralization of Attention Processes in the Human Brain," *Current Opinion in Psychology* 29 (October 2019): 90–96, https://doi.org/10.1016/j.copsyc.2018.12.023.

32. Guido Gainotti, "Introduction," in *Emotions and the Right Side of the Brain*, ed. Guido Gainotti (Cham: Springer International Publishing, 2020), 1–2, https://doi.org/10.1007/978-3-030-34090-2_1.

33. Jeanette Wasserstein and Gerry A. Stefanatos, "The Right Hemisphere and Psychopathology," *Journal of the American Academy of Psychoanalysis* 28, no. 2 (June 2000): 371–395, https://doi.org/10.1521/jaap.1.2000.28.2.371.

34. Sperry, Gazzaniga, and Bogen, "Interhemispheric Relationships."

35. J. E. Bogen, "The Other Side of the Brain: I. Dysgraphia and Dyscopia Following Cerebral Commissurotomy," *Bulletin of the Los Angeles Neurological Society* 34, no. 2 (1969): 73–105.

36. J. E. Bogen, "The Other Side of the Brain: II. An Appositional Mind," *Bulletin of the Los Angeles Neurological Society* 34, no. 3 (July 1969): 135–162.

37. Harrington, *Medicine, Mind, and the Double Brain*, 282.

38. J. E. Bogen and G. M. Bogen, "The Other Side of the Brain: III. The Corpus Callosum and Creativity," *Bulletin of the Los Angeles Neurological Society* 34, no. 4 (October 1969): 191–220.

39. For more on Bogen, see Staub, "The Other Side of the Brain."

40. Dorothy V. M. Bishop, "Cerebral Asymmetry and Language Development: Cause, Correlate, or Consequence?," *Science* 340, no. 6138 (June 14, 2013): 1230531–1230531, https://doi.org/10.1126/science .1230531.

41. Sebastian Ocklenburg et al., "The Ontogenesis of Language Lateralization and Its Relation to Handedness," *Neuroscience & Biobehavioral Reviews* 43 (June 2014): 191–198, https://doi.org/10.1016/j .neubiorev.2014.04.008; Hélène Cochet, "Manual Asymmetries and Hemispheric Specialization: Insight from Developmental Studies," *Neuropsychologia*, The Neural Bases of Hemispheric Specialisation 93 (December 2016): 335–341, https://doi.org/10.1016/j.neuropsychologia .2015.12.019.

42. Michael C. Corballis, "Lateralization of the Human Brain," in *Progress in Brain Research* 195 (2012), 103–121, https://doi.org/10.1016 /B978-0-444-53860-4.00006-4.

43. Staub, "The Other Side of the Brain."

44. Bogen, "The Other Side of the Brain: II," 149–150.

45. K. D. Hoppe, "Hemispheric Specialization and Creativity," *Psychiatric Clinics of North America* 11, no. 3 (September 1988): 303–315.

46. Graham Wallas, *The Art of Thought* (Kent, England: Solis Press, 1926).

47. J. E. Bogen and G. M. Bogen, "Creativity and the Corpus Callosum," *Psychiatric Clinics of North America* 11, no. 3 (September 1988): 299.

48. The notion that complex interactions characterized the operations of the hemispheres was applied at the time to noncreative functions as well. The homeostatic brain model, for instance, held that

although relative hemispheric specialization was certainly the case, there was far more flexibility in lateralization than acknowledged. After all, even while a hemisphere performs a specific function, it continually interacts with the whole brain in sending and receiving information. This implied an internal dynamic network at play that was in service of attaining homeostatic balance: Michael Miran and Esta Miran, "The Evolving of the Homeostatic Brain: Neuropsychological Evidence," in *Individual Differences in Hemispheric Specialization*, ed. A. Glass, NATO ASI Series (Boston: Springer New York, 1987), 331–348, https://doi.org/10.1007/978-1-4684-7899-0_20.

49. Marcel Kinsbourne, "Eye and Head Turning Indicates Cerebral Lateralization," *Science* 176, no. 4034 (May 5, 1972): 539–541, https://doi.org/10.1126/science.176.4034.539.

50. Stevan R. Harnad, "Creativity, Lateral Saccades and the Nondominant Hemisphere," *Perceptual and Motor Skills* 34, no. 2 (April 1972): 653–654, https://doi.org/10.2466/pms.1972.34.2.653; Albert N. Katz, "Creativity and Individual Differences in Asymmetric Cerebral Hemispheric Functioning," *Empirical Studies of the Arts* 1, no. 1 (January 1983): 3–16, https://doi.org/10.2190/8YM6-AM76-UMJX-DQLX; Dwight Hines and Colin Martindale, "Induced Lateral Eye-Movements and Creative and Intellectual Performance," *Perceptual and Motor Skills* 39, no. 1 (August 1974): 153–154, https://doi.org/10.2466/pms.1974.39.1.153.

51. Ann W. Brittain, "Creativity and Hemispheric Functioning: A Second Look at Katz's Data," *Empirical Studies of the Arts* 3, no. 1 (January 1985): 105–107, https://doi.org/10.2190/7NE5-B43T-HC8M-XKAT.

52. Colin Martindale and Dwight Hines, "Creativity and Cortical Activation during Creative, Intellectual and EEG Feedback Tasks," *Biological Psychology* 3, no. 2 (September 1975): 91–100, https://doi.org/10.1016/0301-0511(75)90011-3; Colin Martindale and Nancy Hasenfus, "EEG Differences as a Function of Creativity, Stage of the Creative Process, and Effort to Be Original," *Biological Psychology* 6, no. 3 (April 1978): 157–167, https://doi.org/10.1016/0301-0511(78)90018-2.

53. There many issues of note such as the fact that (1) no assessments were made to test whether the tasks used actually reliably prompted creativity-relevant ideation, and (2) the pattern of brain activity

patterns reported were ambiguous and occasionally conflicting when comparing high and low (and medium) creative groups across all three experiments: Colin Martindale et al., "EEG Alpha Asymmetry and Creativity," *Personality and Individual Differences* 5, no. 1 (January 1984): 77–86, https://doi.org/10.1016/0191-8869(84)90140-5.

54. Colin Martindale, "Biological Bases of Creativity," in *Handbook of Creativity*, ed. R. J. Sternberg (Cambridge: Cambridge University Press, 1999), 137–152.

55. Terence Hines, "The Myth of Right Hemisphere Creativity," *Journal of Creative Behavior* 25, no. 3 (1991): 223–227, https://doi.org /10.1002/j.2162-6057.1991.tb01373.x.

56. M. Miran and E. Miran, "Cerebral Asymmetries: Neuropsychological Measurement and Theoretical Issues," *Biological Psychology* 19, no. 3–4 (December 1984): 295–304; Bogen and Bogen, "Creativity and the Corpus Callosum"; Annukka K. Lindell, "Lateral Thinkers Are Not So Laterally Minded: Hemispheric Asymmetry, Interaction, and Creativity," *Laterality* 16, no. 4 (July 2011): 479–498, https://doi.org /10.1080/1357650X.2010.497813.

57. David Galin, "Implications for Psychiatry of Left and Right Cerebral Specialization: A Neurophysiological Context for Unconscious Processes," *Archives of General Psychiatry* 31, no. 4 (October 1974): 572–583, https://doi.org/10.1001/archpsyc.1974.01760160110022; Klaus D. Hoppe, "Split Brains and Psychoanalysis," *The Psychoanalytic Quarterly* 46, no. 2 (April 1977): 220–244, https://doi.org/10.1080/216740 86.1977.11926798; Robert E. Ornstein, *The Psychology of Consciousness*, A Series of Books in Psychology (San Francisco: W. H. Freeman, 1972).

58. Ernst Kris, *Psychoanalytic Explorations in Art* (New York: International Universities Press, 1952).

59. For alternative viewpoints from the psychoanalytic tradition on the interaction between primary and secondary process cognition as well as the history of the idea, see John R. Suler, "Primary Process Thinking and Creativity," *Psychological Bulletin* 88, no. 1 (1980): 144–165, https://doi.org/10.1037/0033-2909.88.1.144.

60. Martindale, "Biological Bases of Creativity," 145.

61. Examples of behavioral studies using divided visual field paradigms: Ruth A. Atchley, Maureen Keeney, and Curt Burgess, "Cerebral Hemispheric Mechanisms Linking Ambiguous Word Meaning Retrieval and Creativity," *Brain and Cognition* 40, no. 3 (August 1999): 479–499, https://doi.org/10.1006/brcg.1999.1080; Miriam Faust and Michal Lavidor, "Semantically Convergent and Semantically Divergent Priming in the Cerebral Hemispheres: Lexical Decision and Semantic Judgment," *Cognitive Brain Research* 17, no. 3 (October 2003): 585–597, https://doi.org/10.1016/s0926-6410(03)00172-1.

62. Examples of neuroscientific studies using neuroimaging: Dana W. Moore et al., "Hemispheric Connectivity and the Visual-Spatial Divergent-Thinking Component of Creativity," *Brain and Cognition* 70, no. 3 (August 2009): 267–272, https://doi.org/10.1016/j.bandc.2009.02.011; Lisa Aziz-Zadeh, Sook-Lei Liew, and Francesco Dandekar, "Exploring the Neural Correlates of Visual Creativity," *Social Cognitive and Affective Neuroscience* 8, no. 4 (April 2013): 475–480, https://doi.org/10.1093/scan/nss021.

63. For examples of using a combination of behavioral and physiological paradigms, see Kristoffer Carl Aberg, Kimberly C. Doell, and Sophie Schwartz, "The 'Creative Right Brain' Revisited: Individual Creativity and Associative Priming in the Right Hemisphere Relate to Hemispheric Asymmetries in Reward Brain Function," *Cerebral Cortex*, September 20, 2016, https://doi.org/10.1093/cercor/bhw288; Casey E. Turner, Michael E. Hahn, and Ronald T. Kellogg, "Semantic Processing in the Left versus Right Cerebral Hemispheres Following Unilateral Hand Contractions," *Laterality* 22, no. 2 (March 2017): 219–232, https://doi.org/10.1080/1357650X.2016.1154861.

64. David S. Rosen et al., "Dual-Process Contributions to Creativity in Jazz Improvisations: An SPM-EEG Study," *NeuroImage* 213 (June 2020): 116632, https://doi.org/10.1016/j.neuroimage.2020.116632; Aberg, Doell, and Schwartz, "The 'Creative Right Brain' Revisited."

65. Christopher J. Wertz et al., "White Matter Correlates of Creative Cognition in a Normal Cohort," *NeuroImage* 208 (March 2020): 116293, https://doi.org/10.1016/j.neuroimage.2019.116293; Yoed N.

Kenett, David Anaki, and Miriam Faust, "Processing of Unconventional Stimuli Requires the Recruitment of the Non-specialized Hemisphere," *Frontiers in Human Neuroscience* 9 (February 9, 2015), https://doi.org/10.3389/fnhum.2015.00032.

66. Melissa Ellamil et al., "Evaluative and Generative Modes of Thought during the Creative Process," *NeuroImage* 59, no. 2 (January 16, 2012): 1783–1794, https://doi.org/10.1016/j.neuroimage.2011.08.008; Jesus G. Cruz-Garza et al., "Neural Decoding of Expressive Human Movement from Scalp Electroencephalography (EEG)," *Frontiers in Human Neuroscience* 8 (2014): 188, https://doi.org/10.3389/fnhum.2014.00188; I. Carlsson, P. E. Wendt, and J. Risberg, "On the Neurobiology of Creativity: Differences in Frontal Activity between High and Low Creative Subjects," *Neuropsychologia* 38, no. 6 (2000): 873–885.

67. Qunlin Chen et al., "Brain Hemispheric Involvement in Visuospatial and Verbal Divergent Thinking," *NeuroImage* 202 (November 15, 2019): 116065, https://doi.org/10.1016/j.neuroimage.2019.116065.

68. Aziz-Zadeh, Liew, and Dandekar, "Exploring the Neural Correlates of Visual Creativity."

69. Gil Gonen-Yaacovi et al., "Rostral and Caudal Prefrontal Contribution to Creativity: A Meta-Analysis of Functional Imaging Data," *Frontiers in Human Neuroscience* 7 (2013), https://doi.org/10.3389/fnhum.2013.00465; Konstantin M. Mihov, Markus Denzler, and Jens Förster, "Hemispheric Specialization and Creative Thinking: A Meta-Analytic Review of Lateralization of Creativity," *Brain and Cognition* 72, no. 3 (April 2010): 442–448, https://doi.org/10.1016/j.bandc.2009.12.007; Arne Dietrich and Riam Kanso, "A Review of EEG, ERP, and Neuroimaging Studies of Creativity and Insight," *Psychological Bulletin* 136, no. 5 (September 2010): 822–848, https://doi.org/10.1037/a0019749.

70. For a case in point, see Mihov, Denzler, and Förster, "Hemispheric Specialization and Creative Thinking," n63, who refer to Atchley, Keeney, and Burgess, "Cerebral Hemispheric Mechanisms," n60, in their meta-analyses stating that this specific "paper suggests right dominant activation" (446). This does not exactly correspond to

the conclusions of Atchley, Keeney, and Burgess who state: "Results suggest that both the left and right hemispheres contribute to the maintenance of multiple word meanings in highly creative subjects, while less creative subjects show sustained subordinate priming only in the right hemisphere or no sustained subordinate priming. These results support an interactive, collaborative theory of verbal creativity" (479).

71. Dietrich and Kanso, "A Review of EEG, ERP, and Neuroimaging Studies," 833.

72. Stephen M. Fiore and Jonathan W. Schooler, "Right Hemisphere Contributions to Creative Problem Solving: Converging Evidence for Divergent Thinking," in *Right Hemisphere Language Comprehension: Perspectives from Cognitive Neuroscience*, ed. Mark Beeman and Christine Chiarello (Mahwah, NJ: L. Erlbaum Associates, 1998), 349–371.

73. Mark Jung-Beeman et al., "Neural Activity When People Solve Verbal Problems with Insight," *PLoS Biology* 2, no. 4 (April 2004): E97, https://doi.org/10.1371/journal.pbio.0020097.

74. Anna Abraham et al., "Creative Conceptual Expansion: A Combined FMRI Replication and Extension Study to Examine Individual Differences in Creativity," *Neuropsychologia*, May 4, 2018, https://doi.org/10.1016/j.neuropsychologia.2018.05.004.

75. Alice W. Flaherty, "Frontotemporal and Dopaminergic Control of Idea Generation and Creative Drive," *Journal of Comparative Neurology* 493, no. 1 (December 5, 2005): 147–153, https://doi.org/10.1002/cne.20768.

76. Cook, *The Brain Code*, 18.

77. Gesa Hartwigsen, Yoshua Bengio, and Danilo Bzdok, "How Does Hemispheric Specialization Contribute to Human-Defining Cognition?," *Neuron*, May 11, 2021, https://doi.org/10.1016/j.neuron.2021.04.024.

78. Marlene Behrmann and David C. Plaut, "A Vision of Graded Hemispheric Specialization," *Annals of the New York Academy of Sciences* 1359 (November 2015): 30–46, https://doi.org/10.1111/nyas.12833.

79. Stéphanie K. Riès, Nina F. Dronkers, and Robert T. Knight, "Choosing Words: Left Hemisphere, Right Hemisphere, or Both? Perspective on the Lateralization of Word Retrieval," *Annals of the New York Academy of Sciences* 1369, no. 1 (April 2016): 111–131, https://doi.org/10.1111/nyas.12993.

80. Benjamin O. Turner et al., "Hemispheric Lateralization in Reasoning," *Annals of the New York Academy of Sciences* 1359 (November 2015): 47–64, https://doi.org/10.1111/nyas.12940.

81. M. E. Roser and M. S. Gazzaniga, "The Interpreter in Human Psychology," in *Evolution of Nervous Systems* (Elsevier, 2007), 503–508, https://doi.org/10.1016/B0-12-370878-8/00030-6.

82. Vinod Goel, "Indeterminacy Tolerance as a Basis of Hemispheric Asymmetry within Prefrontal Cortex," *Frontiers in Human Neuroscience* 9 (2015): 326, https://doi.org/10.3389/fnhum.2015.00326.

83. Vinod Goel, "Chapter 10—Hemispheric Asymmetry in the Prefrontal Cortex for Complex Cognition," in *The Frontal Lobes*, ed. Mark D'Esposito and Jordan H. Grafman, Handbook of Clinical Neurology, vol. 163 (Elsevier, 2019), 179–196, https://doi.org/10.1016/B978-0-12-804281-6.00010-0.

84. Mark Jung-Beeman, "Bilateral Brain Processes for Comprehending Natural Language," *Trends in Cognitive Sciences* 9, no. 11 (November 2005): 512–518, https://doi.org/10.1016/j.tics.2005.09.009.

85. Elkhonon Goldberg, *Creativity: The Human Brain in the Age of Innovation* (New York: Oxford University Press, 2018).

86. The notion that the right hemisphere processes polysemantic information is also noted in other literatures. See Vadim S. Rotenberg, "The Peculiarity of the Right-Hemisphere Function in Depression: Solving the Paradoxes," *Progress in Neuro-Psychopharmacology and Biological Psychiatry* 28, no. 1 (January 2004): 1–13, https://doi.org/10.1016/S0278-5846(03)00163-5; C. A. Seger et al., "Functional Magnetic Resonance Imaging Evidence for Right-Hemisphere Involvement in Processing Unusual Semantic Relationships," *Neuropsychology* 14, no. 3 (July 2000): 361–369.

87. M. J. Beeman and E. M. Bowden, "The Right Hemisphere Maintains Solution-Related Activation for yet-to-Be-Solved Problems," *Memory & Cognition* 28, no. 7 (October 2000): 1231–1241; Edward M. Bowden and Mark Jung-Beeman, "Aha! Insight Experience Correlates with Solution Activation in the Right Hemisphere," *Psychonomic Bulletin & Review* 10, no. 3 (September 2003): 730–737.

88. A similar distinction was also proposed by Alice W. Flaherty, "Brain Illness and Creativity: Mechanisms and Treatment Risks," *Canadian Journal of Psychiatry. Revue Canadienne De Psychiatrie* 56, no. 3 (March 2011): 132–143. In her view, the right hemisphere was important for novelty detection but the left hemisphere was instrumental to novel action.

89. Goldberg, *Creativity*.

90. So persistent that the definition of the term "right-brained" (as entered in the *American Heritage Dictionary of the English Language*) is to be creative and imaginative. For details, see Michael C. Corballis, "Humanity and the Left Hemisphere: The Story of Half a Brain," *Laterality* 26, no. 1–2 (March 4, 2021): 19–33, https://doi.org/10.1080/1357650X.2020.1782929.

91. Staub, "The Other Side of the Brain," 269; for an opposing view to this claim, see Charles G. Wieder, "The Left-Brain/Right-Brain Model of Mind: Ancient Myth in Modern Garb," *Visual Arts Research* 10, no. 2 (1984): 66–72.

92. Jerome S. Bruner, *On Knowing: Essays for the Left Hand* (Cambridge, MA: Belknap Press: An Imprint of Harvard University Press, 1962).

Chapter 2

1. *Playing with Madness*, BBC Worldwide Ltd. (New York: Films Media Group, 2012).

2. As quoted in Eric Maisel, *Staying Sane in the Arts* (New York: Tarcher, 1992), 7.

3. Seneca, *Peace of Mind: De Tranquillitate Animi*, trans. Aubrey Stewart (CreateSpace Independent Publishing Platform, [1900] 2016), see also https://en.wikisource.org/wiki/Of_Peace_of_Mind#XVII.

4. Stephen A. Diamond, *Anger, Madness, and the Daimonic: The Psychological Genesis of Violence, Evil, and Creativity*, SUNY Series in the Philosophy of Psychology (Albany: State University of New York Press, 1996).

5. Such beliefs are often reported as typifying the Middle Ages: Rick Kemp, *Embodied Acting: What Neuroscience Tells Us about Performance* (London; New York: Routledge, 2012); Leigh Ann Craig, "The History of Madness and Mental Illness in the Middle Ages: Directions and Questions," *History Compass* 12, no. 9 (2014): 729–744, https://doi.org/10.1111/hic3.12187; and Carlos Espí Forcén and Fernando Espí Forcén, "Demonic Possessions and Mental Illness: Discussion of Selected Cases in Late Medieval Hagiographical Literature," *Early Science and Medicine* 19, no. 3 (2014): 258–279, https://doi.org/10.1163/15733823-00193p03. However, such beliefs are still prevalent today in several non-Western cultures: Jörg Haustein, "Embodying the Spirit(s): Pentecostal Demonology and Deliverance Discourse in Ethiopia," *Ethnos* 76, no. 4 (December 2011): 534–552, https://doi.org/10.1080/00141844.2011.598235; Najat Khalifa et al., "Beliefs about Jinn, Black Magic and the Evil Eye among Muslims: Age, Gender and First Language Influences," *International Journal of Culture and Mental Health* 4, no. 1 (June 2011): 68–77, https://doi.org/10.1080/17542863.2010.503051; Frederick M. Smith, "Possession, Embodiment, and Ritual in Mental Health Care in India," *Journal of Ritual Studies* 24, no. 2 (2010): 21–35; E. H. Kua, P. H. Chew, and S. M. Ko, "Spirit Possession and Healing among Chinese Psychiatric Patients," *Acta Psychiatrica Scandinavica* 88, no. 6 (December 1993): 447–450, https://doi.org/10.1111/j.1600-0447.1993.tb03489.x.

6. Plato and Socrates were among the many ancient personages of eminence to attribute guidance and direction of the thought processes to such forces. For a historical overview of the concept of the daemonic, see Angus Nicholls, "Daemonic," in *Oxford Research Encyclopedia of Literature* (Oxford University Press, 2021), https://doi.org

/10.1093/acrefore/9780190201098.013.1118; Darrin M. McMahon, *Divine Fury: A History of Genius* (New York: Basic Books, 2013). Contemporary views that point to the mystical nature of the creative process can be gleaned from first-person accounts in this collection: Brewster Ghiselin, ed., *The Creative Process: A Symposium* (Berkeley; London: University of California Press, 1985).

7. Diamond, *Anger, Madness, and the Daimonic*; see also M. Sperber, "The Daimonic: Freudian, Jungian, and Existential Perspectives," *Journal of Analytical Psychology* 20, no. 1 (1975): 41–49, https://doi.org /10.1111/j.1465-5922.1975.00041.x.

8. As cited in Andrew Robinson, *Genius: A Very Short Introduction* (Oxford; New York: Oxford University Press, 2011), 64.

9. Andrew Steptoe, "Artistic Temperament in the Italian Renaissance: A Study of Giorgio Vasari's Lives," in *Genius and the Mind: Studies of Creativity and Temperament*, ed. Andrew Steptoe (Oxford: Oxford University Press, 1998), https://doi.org/10.1093/acprof:oso /9780198523734.001.0001.

10. Robinson, *Genius*, 59.

11. For an examination of how optimal social systems in Renaissance Florence are tied to creative flourishing in that period, see Mihaly Csikszentmihalyi, "Society, Culture, and Person: A Systems View of Creativity," in *The Systems Model of Creativity: The Collected Works of Mihaly Csikszentmihalyi*, ed. Mihaly Csikszentmihalyi (Dordrecht: Springer Netherlands, 2014), 47–61, https://doi.org/10.1007 /978-94-017-9085-7_4.

12. For a brief and excellent overview, see Roy Porter, *Madness: A Brief History* (Oxford; New York: Oxford University Press, 2002); for a comprehensive analysis, see Petteri Pietikainen, *Madness: A History* (London: Routledge, 2015), https://doi.org/10.4324/9781315708966.

13. Miguel A. Faria, "Violence, Mental Illness, and the Brain—A Brief History of Psychosurgery: Part 1—From Trephination to Lobotomy," *Surgical Neurology International* 4 (April 5, 2013): 49, https://www.ncbi .nlm.nih.gov/pmc/articles/PMC3640229/.

14. Porter, *Madness*, 177–182; for an early example, see John Haslam, *Illustrations of Madness: Exhibiting a Singular Case of Insanity and a No Less Remarkable Difference in Medical Opinion . . . with a Description of the Tortures Experienced [by the Patient, James Tilly Matthews, in Hallucinations]* (London: G. Hayden for Rivingtons, etc., 1810), https://wellcomecollection.org/works/e6n82yw6.

15. John Locke, *An Essay Concerning Humane Understanding* (London: Printed by Eliz. Holt for Thomas Basset, 1690), https://quod.lib.umich.edu/e/eebo/A48874.0001.001?view=toc.

16. Porter, *Madness*, 66.

17. From John Dryden, Absalom and Achitophel (1681) pt. 1, I. 163, as stated in Susan Ratcliffe, *Oxford Essential Quotations*, 5th ed. (Oxford University Press, 2017), https://www.oxfordreference.com/display/10.1093/acref/9780191843730.001.0001/q-oro-ed5-00003835.

18. R. D. Laing, *The Politics of Experience and the Bird of Paradise* (Harmondsworth: Penguin, 1967).

19. John Ferguson Nisbet, *The Insanity of Genius and the General Inequality of Human Faculty, Physiologically Considered* (London: Grant Richards, 1900), https://wellcomecollection.org/works/fppyumf7.

20. Cesar Lombroso met Lev Tolstoy in the hope of verifying his views of degeneration in relation to genius. The encounter famously did not go to plan. Tolstoy exclaimed, "All this is nonsense!" on hearing these views and described Lombroso as "an ingenuous and limited old man": Paolo Mazzarello, "Cesare Lombroso: An Anthropologist between Evolution and Degeneration," *Functional Neurology* 26, no. 2 (July 3, 2011): 97–101.

21. Robert A. Power et al., "Polygenic Risk Scores for Schizophrenia and Bipolar Disorder Predict Creativity," *Nature Neuroscience* 18, no. 7 (July 2015): 953–955, https://doi.org/10.1038/nn.4040.

22. Cesare Lombroso, *The Man of Genius* (London: Walter Scott, 1891), https://www.gutenberg.org/ebooks/50539.

23. The antipsychiatry movement held that mental illness was a fabrication by the psychiatric establishment: Thomas Stephen Szasz, *The*

Myth of Mental Illness: Foundations of a Theory of Personal Conduct, 2nd rev. ed. (New York; Cambridge; Philadelphia [etc.]: Harper & Row, 1974). The need to factor in power dynamics was also proposed in other highly influential considerations of madness as a sociocultural construct; see, for example, Michel Foucault, *Madness and Civilization: A History of Insanity in the Age of Reason* (New York: Vintage Books, 1961).

24. Martin Roth, *The Reality of Mental Illness* (Cambridge; New York: Cambridge University Press, 1986).

25. Richard Alfred Hunter and Ida Macalpine, *Three Hundred Years of Psychiatry, 1535–1860: A History Presented in Selected English Texts* (Hartsdale, NY: Carlisle Publishing, 1963).

26. Peter Tyrer, "A Comparison of DSM and ICD Classifications of Mental Disorder," *Advances in Psychiatric Treatment* 20, no. 4 (July 2014): 280–285, https://doi.org/10.1192/apt.bp.113.011296.

27. The case for valuing the term "madness" and not substituting it across all instances with "mental illness" (which is tantamount to only lending a medical lens to the state and, in doing so, ignoring the wider sociocultural, political, and economic contexts in which such states occur) is made in this paper: Tomi Gomory, David Cohen, and Stuart A. Kirk, "Madness or Mental Illness? Revisiting Historians of Psychiatry," *Current Psychology* 32, no. 2 (June 2013): 119–135, https://doi.org/10.1007/s12144-013-9168-3.

28. American Psychiatric Association, ed., *Diagnostic and Statistical Manual of Mental Disorders: DSM-5*, 5th ed. (Washington, DC: American Psychiatric Association, 2013).

29. Deanna M. Barch et al., "Logic and Justification for Dimensional Assessment of Symptoms and Related Clinical Phenomena in Psychosis: Relevance to DSM-5," *Schizophrenia Research*, DSM-5, 150, no. 1 (October 2013): 15–20, https://doi.org/10.1016/j.schres.2013.04.027.

30. National Institute of Mental Health, "Schizophrenia," accessed October 31, 2021, https://www.nimh.nih.gov/health/topics/schizophrenia.

31. National Institute of Mental Health, "Bipolar Disorder," accessed October 31, 2021, https://www.nimh.nih.gov/health/topics/bipolar -disorder.

32. Dean Keith Simonton, "The Mad (Creative) Genius: What Do We Know after a Century of Historiometric Research?," in *Creativity and Mental Illness* (New York: Cambridge University Press, 2014), 25–41, https://doi.org/10.1017/CBO9781139128902.004.

33. N. C. Andreasen, "Creativity and Mental Illness: Prevalence Rates in Writers and Their First-Degree Relatives," *American Journal of Psychiatry* 144, no. 10 (October 1987): 1288–1292; K. R. Jamison, "Mood Disorders and Patterns of Creativity in British Writers and Artists," *Psychiatry* 52, no. 2 (May 1989): 125–134.

34. A. M. Ludwig, *The Price of Greatness: Resolving the Creativity and Madness Controversy* (New York: Guilford Press, 1995), 4–5.

35. George Becker, "The Association of Creativity and Psychopathology: Its Cultural-Historical Origins," *Creativity Research Journal* 13, no. 1 (January 2001): 45–53, https://doi.org/10.1207/S15326934CRJ1301_6.

36. Adele Juda, "The Relationship between Highest Mental Capacity and Psychic Abnormalities," *American Journal of Psychiatry* 106, no. 4 (October 1949): 296–307, https://doi.org/10.1176/ajp.106.4.296; Ruth L. Richards, "Relationships between Creativity and Psychopathology: An Evaluation and Interpretation of the Evidence," *Genetic Psychology Monographs* 103, no. 2 (1981): 261–324.

37. For arguments in favor of keeping the disorders separate, see Heinz Grunze and Marcelo Cetkovich-Bakmas, "'Apples and Pears Are Similar, but Still Different Things.' Bipolar Disorder and Schizophrenia—Discrete Disorders or Just Dimensions?," *Journal of Affective Disorders* 290 (July 2021): 178–187, https://doi.org/10.1016/j.jad.2021 .04.064.

38. Reviews that make this argument include the following: Godfrey D. Pearlson, "Etiologic, Phenomenologic, and Endophenotypic Overlap of Schizophrenia and Bipolar Disorder," *Annual Review of Clinical Psychology* 11 (2015): 251–281, https://doi.org/10.1146/annurev-clinpsy -032814-112915; Wade H. Berrettini, "Susceptibility Loci for Bipolar

Disorder: Overlap with Inherited Vulnerability to Schizophrenia,"
Biological Psychiatry 47, no. 3 (February 2000): 245–251, https://doi
.org/10.1016/s0006-3223(99)00226-7; Ellen Ji et al., "From the Micro-
scope to the Magnet: Disconnection in Schizophrenia and Bipolar
Disorder," *Neuroscience & Biobehavioral Reviews* 98 (March 2019):
47–57, https://doi.org/10.1016/j.neubiorev.2019.01.005; Gunvant K.
Thaker, "Neurophysiological Endophenotypes across Bipolar and
Schizophrenia Psychosis," *Schizophrenia Bulletin* 34, no. 4 (July 2008):
760–773, https://doi.org/10.1093/schbul/sbn049; Berna Yalincetin et
al., "Formal Thought Disorder in Schizophrenia and Bipolar Disorder:
A Systematic Review and Meta-Analysis," *Schizophrenia Research* 185
(July 2017): 2–8, https://doi.org/10.1016/j.schres.2016.12.015.

39. In the context of creativity, the work of Hans Eysenck is especially
relevant. He argued for a common "psychoticism" dimension and
advocated for the positive link between creativity and psychoticism:
Hans J. Eysenck, *Genius: The Natural History of Creativity* (Cambridge,
MA: Cambridge University Press, 1995).

40. Yuji Yamada et al., "Specificity and Continuity of Schizophrenia
and Bipolar Disorder: Relation to Biomarkers," *Current Pharmaceutical
Design* 26, no. 2 (2020): 191–200, https://doi.org/10.2174/138161282
5666191216153508.

41. Simon Kyaga et al., "Creativity and Mental Disorder: Family
Study of 300,000 People with Severe Mental Disorder," *British Journal
of Psychiatry* 199, no. 5 (November 2011): 373–379, https://doi.org
/10.1192/bjp.bp.110.085316.

42. The effect is modest given the odds ratio of 1.08 for this association: Simon Kyaga et al., "Mental Illness, Suicide and Creativity:
40-Year Prospective Total Population Study," *Journal of Psychiatric
Research* 47, no. 1 (January 2013): 83–90, https://doi.org/10.1016/j
.jpsychires.2012.09.010. For a good explanation on how to interpret
odds ratios, see Magdalena Szumilas, "Explaining Odds Ratios," *Journal of the Canadian Academy of Child and Adolescent Psychiatry* 19, no.
3 (August 2010): 227–229.

43. Felix Post, "Creativity and Psychopathology: A Study of 291
World-Famous Men," *British Journal of Psychiatry* 165, no. 1 (July

1994): 22–34, http://dx.doi.org/10.1192/bjp.165.1.22; see also Felix Post, "Verbal Creativity, Depression and Alcoholism: An Investigation of One Hundred American and British Writers," *British Journal of Psychiatry* 168, no. 5 (May 1996): 545–555.

44. E. Thys, B. Sabbe, and M. De Hert, "The Assessment of Creativity in Creativity/Psychopathology Research—A Systematic Review," *Cognitive Neuropsychiatry* 19, no. 4 (July 4, 2014): 359–377, https://doi.org/10.1080/13546805.2013.877384.

45. Márcio Gerhardt Soeiro-de-Souza et al., "Creativity and Executive Function across Manic, Mixed and Depressive Episodes in Bipolar I Disorder," *Journal of Affective Disorders* 135, no. 1–3 (December 2011): 292–297, https://doi.org/10.1016/j.jad.2011.06.024.

46. Anna Abraham et al., "Creative Thinking in Schizophrenia: The Role of Executive Dysfunction and Symptom Severity," *Cognitive Neuropsychiatry* 12, no. 3 (May 2007): 235–258, https://doi.org/10.1080/13546800601046714.

47. Eva Burkhardt et al., "Creativity in Persons At-Risk for Bipolar Disorder—A Pilot Study," *Early Intervention in Psychiatry* 13, no. 5 (2019): 1165–1172, https://doi.org/10.1111/eip.12748.

48. Selcuk Acar, Xiao Chen, and Nur Cayirdag, "Schizophrenia and Creativity: A Meta-Analytic Review," *Schizophrenia Research* 195 (May 2018): 23–31, https://doi.org/10.1016/j.schres.2017.08.036; Christa L. Taylor, "Creativity and Mood Disorder: A Systematic Review and Meta-Analysis," *Perspectives on Psychological Science: A Journal of the Association for Psychological Science* 12, no. 6 (November 2017): 1040–1076, https://doi.org/10.1177/1745691617699653.

49. This fits with views of the "mythconception" of the madness-creativity link: Schlesinger, "Creative Mythconceptions"; Dietrich, "The Mythconception of the Mad Genius."

50. Ruth Richards was an early advocate of the dimensional approach in coming to terms with the madness-creativity link, and many of her insights have been borne out and consolidated by later work: Ruth Richards et al., "Creativity in Manic-Depressives, Cyclothymes, Their Normal Relatives, and Control Subjects," *Journal of Abnormal Psychol-*

ogy 97, no. 3 (1988): 281–288, https://doi.org/10.1037/0021-843X.97 .3.281; Ruth Richards, "Creativity and the Schizophrenia Spectrum: More and More Interesting," *Creativity Research Journal* 13, no. 1 (2001): 111–132, https://doi.org/10.1207/S15326934CRJ1301_13.

51. Christine Mohr and Gordon Claridge, "Schizotypy—Do Not Worry, It Is Not All Worrisome," *Schizophrenia Bulletin* 41 Suppl. 2 (March 2015): S436–443, https://doi.org/10.1093/schbul/sbu185.

52. Odds ratios > 1.2: Kyaga et al., "Mental Illness, Suicide and Creativity."

53. Matthijs Baas et al., "Mad Genius Revisited: Vulnerability to Psychopathology, Biobehavioral Approach-Avoidance, and Creativity," *Psychological Bulletin* 142, no. 6 (June 2016): 668–692, https://doi.org /10.1037/bul0000049; Selcuk Acar and Sedat Sen, "A Multilevel Meta-analysis of the Relationship between Creativity and Schizotypy," *Psychology of Aesthetics, Creativity, and the Arts* 7, no. 3 (August 2013): 214–228, https://doi.org/10.1037/a0031975.

54. These include the Barron-Welsh Art Scale, divergent thinking tests, and creative cognition measures: Burkhardt et al., "Creativity in Persons At-Risk for Bipolar Disorder"; H. J. Eysenck and A. Furnham, "Personality and the Barron-Welsh Art Scale," *Perceptual and Motor Skills* 76, no. 3 (June 1993): 837–838, https://doi.org/10.2466/pms .1993.76.3.837; David Schuldberg et al., "Creativity and Schizotypal Traits: Creativity Test Scores and Perceptual Aberration, Magical Ideation, and Impulsive Nonconformity," *Journal of Nervous and Mental Disease* 176, no. 11 (November 1988): 648–657, https://doi.org/10 .1097/00005053-198811000-00002; Anna Abraham et al., "Conceptual Expansion and Creative Imagery as a Function of Psychoticism," *Consciousness and Cognition* 14, no. 3 (September 2005): 520–534, https://doi.org/10.1016/j.concog.2004.12.003; Anna Abraham and Sabine Windmann, "Selective Information Processing Advantages in Creative Cognition as a Function of Schizotypy," *Creativity Research Journal* 20, no. 1 (2008): 1–6, https://doi.org/10.1080/10400410701 839819.

55. Matthijs Baas et al., "Vulnerability to Psychopathology and Creativity: The Role of Approach-Avoidance Motivation and Novelty

Seeking," *Psychology of Aesthetics, Creativity, and the Arts 14*, no. 3 (2020): 334–352. Commonalities between bipolar disorder and positive or disorganized traits of schizotypy are also advocated by many: Franck Schürhoff et al., "Schizotypal Dimensions: Continuity between Schizophrenia and Bipolar Disorders," *Schizophrenia Research* 80, no. 2–3 (December 15, 2005): 235–242, https://doi.org/10.1016/j .schres.2005.07.009; Janusz K. Rybakowski and Paulina Klonowska, "Bipolar Mood Disorder, Creativity and Schizotypy: An Experimental Study," *Psychopathology* 44, no. 5 (2011): 296–302, https://doi.org/10 .1159/000322814.

56. Małgorzata A. Gocłowska et al., "Novelty Seeking Is Linked to Openness and Extraversion, and Can Lead to Greater Creative Performance," *Journal of Personality* 87, no. 2 (2019): 252–266, https://doi .org/10.1111/jopy.12387.

57. Giles St J. Burch et al., "Schizotypy and Creativity in Visual Artists," *British Journal of Psychology* 97, no. 2 (May 2006): 177–190, https://doi.org/10.1348/000712605X60030; David Rawlings and Ann Locarnini, "Dimensional Schizotypy, Autism, and Unusual Word Associations in Artists and Scientists," *Journal of Research in Personality* 42, no. 2 (April 2008): 465–471, https://doi.org/10.1016/j.jrp.2007.06 .005; Daniel Nettle, "Schizotypy and Mental Health amongst Poets, Visual Artists, and Mathematicians," *Journal of Research in Personality* 40, no. 6 (December 2006): 876–890, https://doi.org/10.1016/j.jrp .2005.09.004; Crystal Gibson, Bradley S. Folley, and Sohee Park, "Enhanced Divergent Thinking and Creativity in Musicians: A Behavioral and Near-Infrared Spectroscopy Study," *Brain and Cognition* 69, no. 1 (February 2009): 162–169, https://doi.org/10.1016/j.bandc.2008 .07.009.

58. Nicola J. Holt, "The Expression of Schizotypy in the Daily Lives of Artists," *Psychology of Aesthetics, Creativity, and the Arts* 13, no. 3 (2019): 359, https://doi.org/10.1037/aca0000176.

59. Katherine Taylor, I. Fletcher, and F. Lobban, "Exploring the Links between the Phenomenology of Creativity and Bipolar Disorder," *Journal of Affective Disorders* 174 (March 15, 2015): 658–664, https:// doi.org/10.1016/j.jad.2014.10.040.

60. B. Nelson and D. Rawlings, "Relating Schizotypy and Personality to the Phenomenology of Creativity," *Schizophrenia Bulletin* 36, no. 2 (March 2010): 388–399, https://doi.org/10.1093/schbul/sbn098; Holt, "The Expression of Schizotypy in the Daily Lives of Artists."

61. Andreas Fink et al., "Creativity: Genius, Madness, or a Combination of Both?," *Psychology of Aesthetics, Creativity, and the Arts* 6, no. 1 (2012): 11–18, https://doi.org/10.1037/a0024874; Lixia Wang et al., "High Schizotypal Individuals Are More Creative? The Mediation Roles of Overinclusive Thinking and Cognitive Inhibition," *Frontiers in Psychology* 9 (September 21, 2018): 1766, https://doi.org/10.3389/fpsyg.2018.01766.

62. For a brief review, see Andreas Fink et al., "Creativity and Psychopathology: Are There Similar Mental Processes Involved in Creativity and in Psychosis-Proneness?," *Frontiers in Psychology* 5 (2014): 1211, https://doi.org/10.3389/fpsyg.2014.01211.

63. Recent influential accounts include: Shelley H. Carson, "Creativity and Psychopathology: A Shared Vulnerability Model," *Canadian Journal of Psychiatry. Revue Canadienne De Psychiatrie* 56, no. 3 (March 2011): 144–153; Tiffany A. Greenwood, "Creativity and Bipolar Disorder: A Shared Genetic Vulnerability," *Annual Review of Clinical Psychology* 16 (May 7, 2020): 239–264, https://doi.org/10.1146/annurev-clinpsy-050718-095449.

64. Jón L.Karlsson, "Genetic Association of Giftedness and Creativity with Schizophrenia," *Hereditas* 66, no. 2 (1970): 177–182, https://doi.org/10.1111/j.1601-5223.1970.tb02343.x; Daniel Nettle, *Strong Imagination: Madness, Creativity and Human Nature* (Oxford; New York: Oxford University Press, 2001).

65. Dennis K. Kinney and Ruth Richards, "Creativity as 'Compensatory Advantage': Bipolar and Schizophrenic Liability, the Inverted-U Hypothesis, and Practical Implications," in *Creativity and Mental Illness* (New York: Cambridge University Press, 2014), 295–317, https://doi.org/10.1017/CBO9781139128902.019; David Schuldberg, "Six Subclinical Spectrum Traits in Normal Creativity," *Creativity Research Journal* 13, no. 1 (January 2001): 5–16, https://doi.org/10.1207/S15326934CRJ1301_2; Anna Abraham, "Is There an Inverted-U Relationship

between Creativity and Psychopathology?," *Frontiers in Psychology* 5 (2014): 750, https://doi.org/10.3389/fpsyg.2014.00750.

66. Kyaga et al., "Mental Illness, Suicide and Creativity."

67. This article explores the possibility that isolation at multiple levels of the creative process is what makes writers particularly susceptible to mental illness: Anna Abraham, "Creativity or Creativities? Why Context Matters," *Design Studies* 78 (January 2022): 101060, https://doi.org/10.1016/j.destud.2021.101060.

68. Odds ratios range between 1.3–1.9: J. H. MacCabe et al., "Artistic Creativity and Risk for Schizophrenia, Bipolar Disorder and Unipolar Depression: A Swedish Population-Based Case-Control Study and Sib-Pair Analysis," *British Journal of Psychiatry* 212, no. 6 (June 2018): 370–376, https://doi.org/10.1192/bjp.2018.23.

69. For an examination of contextual factors in creativity, see Abraham, "Creativity or Creativities? Why Context Matters"; Teresa M. Amabile, *Creativity in Context* (Boulder, CO: Westview Press, 1996); Mihaly Csikszentmihalyi, "Society, Culture, and Person: A Systems View of Creativity," in *The Systems Model of Creativity: The Collected Works of Mihaly Csikszentmihalyi*, ed. Mihaly Csikszentmihalyi (Dordrecht: Springer Netherlands, 2014), 47–61, https://doi.org/10.1007/978-94-017-9085-7_4.

70. National Endowment for the Arts, "Artists and Other Cultural Workers: A Statistical Portrait," April 2019, https://www.arts.gov/sites/default/files/Artists_and_Other_Cultural_Workers.pdf.

71. Law Commission of Ontario, "V. What Types of Jobs Are Precarious?," accessed November 8, 2021, https://www.lco-cdo.org/en/our-current-projects/vulnerable-workers-and-precarious-work/commissioned-papers/precarious-jobs-in-ontario-mapping-dimensions-of-labour-market-insecurity-by-workers-social-location-and-context/v-what-types-of-jobs-are-precarious.

72. James V. Marrone, Susan A. Resetar, and Daniel Schwam, "The Pandemic Is a Disaster for Artists," *The RAND Blog* (blog), August 4, 2020, https://www.rand.org/blog/2020/07/the-pandemic-is-a-disaster-for-artists.html.

73. These issues also affect the study of the creative brain: Anna Abraham, "The Promises and Perils of the Neuroscience of Creativity," *Frontiers in Human Neuroscience* 7: (2013): 246, https://doi.org/10.3389/fnhum.2013.00246.

74. This is wonderfully explored in a TED Talk by Elizabeth Gilbert, *Your Elusive Creative Genius*, 2009, https://www.ted.com/talks/elizabeth_gilbert_your_elusive_creative_genius.

75. Albert Rothenberg, *Creativity and Madness: New Findings and Old Stereotypes* (Baltimore, MD: Johns Hopkins University Press, 1990), 36.

76. This is examined in Abraham, "Creativity or Creativities? Why Context Matters."

77. The author explores how romanticism's embrace of the mad poet notion was founded as an oppositional stance toward modernity, and how without the romanticization of madness in that era there would probably be no surrealism, Dadaism, and further artistic legacies that followed in a similar vein of challenging the status quo: James Whitehead, *Madness and the Romantic Poet: A Critical History* (Oxford: Oxford University Press, 2017), 210.

78. Pietikainen, *Madness*, 10.

79. Shakespeare, William, 1564–1616, *A Midsummer Night's Dream* (New York: Signet Classic, 1998), act 5, scene 1, lines 4–8.

80. Rothenberg, *Creativity and Madness*, 35.

81. Anthony Storr, *The Dynamics of Creation*, 1st Ballantine Books ed. (New York: Ballantine Books, 1993), 294.

82. Geoffrey I. Wills, "Forty Lives in the Bebop Business: Mental Health in a Group of Eminent Jazz Musicians," *British Journal of Psychiatry* 183 (September 2003): 255–259, https://doi.org/10.1192/bjp.183.3.255; Taylor L. Benson and Sohee Park, "Exceptional Visuospatial Imagery in Schizophrenia; Implications for Madness and Creativity," *Frontiers in Human Neuroscience* 7 (2013): 756, https://doi.org/10.3389/fnhum.2013.00756; Orjan de Manzano et al., "Thinking Outside a Less Intact Box: Thalamic Dopamine D2 Receptor Densities Are Negatively Related to Psychometric Creativity in Healthy Indi-

viduals," *PloS ONE* 5, no. 5 (2010): e10670, https://doi.org/10.1371 /journal.pone.0010670; N. J. Andreasen and Pauline S. Powers, "Overinclusive Thinking in Mania and Schizophrenia," *British Journal of Psychiatry* 125 (1974): 452–456, https://doi.org/10.1192/bjp.125.5 .452; Hans J. Eysenck, "Creativity and Personality: Suggestions for a Theory," *Psychological Inquiry* 4, no. 3 (July 1993): 147–178, https:// doi.org/10.1207/s15327965pli0403_1.

83. Stein, "Creativity and Culture," 311.

84. Frank Barron, *Creativity and Psychological Health* (Oxford: D. Van Nostrand, 1963), 234.

85. The wonderful phrase "controlled weirdness" was used by an inventor when describing his thought process to Barron: Frank Barron, "Controllable Oddness as a Resource in Creativity," *Psychological Inquiry* 4, no. 3 (1993): 183.

86. Clifford A. Pickover, *Strange Brains and Genius: The Secret Lives of Eccentric Scientists and Madmen* (New York: Plenum Trade, 1998).

Chapter 3

1. Invaluable reflections on his experimentations with mescal. The paper also details postulations about the factors that play a role in explaining individual differences in such experiences, as well as the strengths and limitations of such mind-expanding experiences: Havelock Ellis, "Mescal: A New Artificial Paradise," *Contemporary Review* 73 (1898): 130–141.

2. Excerpt from an interview with the artist (228) (author's translation): Richard P. Hartmann, *Malerei aus Bereichen des Unbewussten: Künstler experimentieren unter LSD* (Köln: M. DuMont Schauberg, 1974).

3. The focus in this chapter is limited to hallucinogenic drugs. For an overview on the class of drugs, see NIDA, "DrugFacts: Hallucinogens," April 22, 2019, https://nida.nih.gov/sites/default/files/drugfacts -hallucinogens.pdf. For a brief overview of the impact of other drugs and related substances on creativity, see Iain Smith, "Psychostimulants

and Artistic, Musical, and Literary Creativity," *International Review of Neurobiology* 120 (2015): 301–326, https://doi.org/10.1016/bs.irn.2015.04.001.

4. The first formal study of the impact of a psychoactive drug on the central nervous system: Jacques Joseph Moreau, *Du hachisch et de l'aliénation mentale: études psychologiques* (Paris: Fortin, Masson, 1845), http://archive.org/details/duhachischetdela00more.

5. Fascinating examples from the prehistoric era are documented in Giorgio Samorini, "The Oldest Representations of Hallucinogenic Mushrooms in the World (Sahara Desert, 9000–7000 B.P.)," *Integration* 2, no. 3 (1992): 69–78; and Brian P. Akers et al., "A Prehistoric Mural in Spain Depicting Neurotropic Psilocybe Mushrooms?," *Economic Botany* 65, no. 2 (June 2011): 121–128, https://doi.org/10.1007/s12231-011-9152-5.

6. Influential works from the modern era include James Fadiman, *The Psychedelic Explorer's Guide: Safe, Therapeutic, and Sacred Journeys*, illustrated ed. (Rochester, VT: Park Street Press, 2011); Albert Hofmann, *LSD, My Problem Child* (New York: McGraw-Hill, 1980); and Aldous Huxley, *The Doors of Perception and Heaven and Hell*, 1st Perennial Classics ed. (New York: Perennial Classics, 2004).

7. A recent example of how findings of a neuroscientific study on LSD (that had yet to be peer reviewed) are broadcast in a major news outlet with a mere two days between the study preprint being posted in an open repository (May 17, 2021) and the publication of the media article (May 19, 2021): Ian Sample, "Acid Test: Scientists Show How LSD Opens Doors of Perception," *Guardian*, May 19, 2021, sec. Science, http://www.theguardian.com/science/2021/may/19/acid-test-scientists-show-how-lsd-opens-doors-of-perception.

8. It is worth exploring older classic studies that sought to address such questions, such as William McGlothlin, Sidney Cohen, and Marcella S. McGlothlin, "Long Lasting Effects of LSD on Normals," *Archives of General Psychiatry* 17, no. 5 (November 1967): 521–532, https://doi.org/10.1001/archpsyc.1967.01730290009002, as well as contemporary studies that do so, such as Suzanne L. Russ et al.,

"States and Traits Related to the Quality and Consequences of Psyche-delic Experiences.," *Psychology of Consciousness: Theory, Research, and Practice* 6, no. 1 (March 2019): 1–21, https://doi.org/10.1037/cns 0000169.

9. Evidence showing long-term effects two months later: R. R. Griffiths et al., "Psilocybin Can Occasion Mystical-Type Experiences Having Substantial and Sustained Personal Meaning and Spiritual Significance," *Psychopharmacology* 187, no. 3 (August 2006): 268–283, https://doi.org/10.1007/s00213-006-0457-5; and twelve months later: Yasmin Schmid and Matthias E. Liechti, "Long-Lasting Subjective Effects of LSD in Normal Subjects," *Psychopharmacology* 235, no. 2 (February 2018): 535–545, https://doi.org/10.1007/s00213-017-4733-3.

10. Roland R. Griffiths et al., "Survey of Subjective 'God Encounter Experiences': Comparisons among Naturally Occurring Experiences and Those Occasioned by the Classic Psychedelics Psilocybin, LSD, Ayahuasca, or DMT," ed. Rosemary Frey, *PLoS ONE* 14, no. 4 (April 23, 2019): e0214377, https://doi.org/10.1371/journal.pone.0214377.

11. Selected examples include Harri Nyberg, "Religious Use of Hallu-cinogenic Fungi: A Comparison between Siberian and Mesoamerican Cultures," *Karstenia* 32, no. 2 (1992): 71–80, https://doi.org/10.29203 /ka.1992.294; and Jean-Francois Sobiecki, "A Review of Plants Used in Divination in Southern Africa and Their Psychoactive Effects," *South-ern African Humanities* 20 (2008): 1–19.

12. Tomislav Majić, Timo T. Schmidt, and Jürgen Gallinat, "Peak Experiences and the Afterglow Phenomenon: When and How Do Therapeutic Effects of Hallucinogens Depend on Psychedelic Expe-riences?," *Journal of Psychopharmacology* (Oxford) 29, no. 3 (March 2015): 241–253, https://doi.org/10.1177/0269881114568040.

13. Frederick S. Barrett et al., "Emotions and Brain Function Are Altered up to One Month after a Single High Dose of Psilocybin," *Scientific Reports* 10, no. 1 (February 10, 2020): 2214, https://doi.org /10.1038/s41598-020-59282-y; Martin Korsbak Madsen et al., "A Single Psilocybin Dose Is Associated with Long-Term Increased Mind-fulness, Preceded by a Proportional Change in Neocortical 5-HT2A

Receptor Binding," *European Neuropsychopharmacology* 33 (April 2020): 71–80, https://doi.org/10.1016/j.euroneuro.2020.02.001.

14. Insights on the impact of repeated use of psychedelics on creativity largely stem from qualitative studies on microdosing. See Fadiman, *The Psychedelic Explorer's Guide*; Anita Hardon, "Chemical Creativity," in *Chemical Youth: Navigating Uncertainty in Search of the Good Life*, ed. Anita Hardon, Critical Studies in Risk and Uncertainty (Cham: Springer International Publishing, 2021), 247–279, https://doi.org/10.1007/978-3-030-57081-1_8; Kim P. C. Kuypers et al., "Microdosing Psychedelics: More Questions than Answers? An Overview and Suggestions for Future Research," *Journal of Psychopharmacology* 33, no. 9 (September 2019): 1039–1057, https://doi.org/10.1177/0269881119857204.

15. The Geneplore model: R. A. Finke, T. B. Ward, and S. M. Smith, *Creative Cognition: Theory, Research, and Applications* (Cambridge, MA: MIT Press, 1996).

16. Drawn from Henri Poincaré's reflections on this creative process: Wallas, *The Art of Thought*.

17. James C. Kaufman, "The Creative Construct," *RSA Journal* 162, no. 5565 (2016): 26.

18. For an overview of processes relevant to creativity, see Anna Abraham, *The Neuroscience of Creativity* (New York: Cambridge University Press, 2018).

19. Majić, Schmidt, and Gallinat, "Peak Experiences and the Afterglow Phenomenon," 3.

20. Juliette Bowers, "Flow and Peak Experiences," in *Handbook of Medical and Psychological Hypnosis: Foundations, Applications, and Professional Issues* (New York: Springer Publishing Company, 2017), 559–563.

21. Oliver Stoll and Jan M. Pithan, "Running and Flow: Does Controlled Running Lead to Flow-States? Testing the Transient Hypofontality Theory," in *Flow Experience: Empirical Research and Applications*,

ed. László Harmat et al. (Cham: Springer International Publishing, 2016), 65–75, https://doi.org/10.1007/978-3-319-28634-1_5.

22. For selected examples, see Simon Høffding, *A Phenomenology of Musical Absorption*, New Directions in Philosophy and Cognitive Science (Cham, Switzerland: Palgrave Macmillan, 2018), https://doi.org/10.1007/978-3-030-00659-4; Joseph Glicksohn and Aviva Berkovich-Ohana, "Absorption, Immersion, and Consciousness," in *Video Game Play and Consciousness*, Perspectives on Cognitive Psychology (Hauppauge, NY: Nova Science Publishers, 2012), 83–99; Anissa Rivers et al., "Empathic Features and Absorption in Fantasy Role-Playing," *American Journal of Clinical Hypnosis* 58, no. 3 (January 12, 2016): 286–294, https://doi.org/10.1080/00029157.2015.1103696; Stoney Brooks and Phil Longstreet, "Social Networking's Peril: Cognitive Absorption, Social Networking Usage, and Depression," *Cyberpsychology: Journal of Psychosocial Research on Cyberspace* 9, no. 4 (December 2015), https://doi.org/10.5817/CP2015-4-5.

23. A selection of three contemporary studies that examine this relation: Fruzsina Iszáj, Mark D. Griffiths, and Zsolt Demetrovics, "Creativity and Psychoactive Substance Use: A Systematic Review," *International Journal of Mental Health and Addiction* 15, no. 5 (October 2017): 1135–1149, https://doi.org/10.1007/s11469-016-9709-8; Fruzsina Iszáj et al., "Substance Use and Psychological Disorders among Art and Non-art University Students: An Empirical Self-Report Survey," *International Journal of Mental Health and Addiction* 16, no. 1 (2018): 125–135, https://doi.org/10.1007/s11469-017-9812-5; Antonio Preti and Marcello Vellante, "Creativity and Psychopathology: Higher Rates of Psychosis Proneness and Nonright-Handedness among Creative Artists Compared to Same Age and Gender Peers," *Journal of Nervous and Mental Disease* 195, no. 10 (October 2007): 837–845, https://doi.org/10.1097/NMD.0b013e3181568180.

24. Papers by Jos ten Berge provide engaging qualitative examinations of such expectancies: Jos ten Berge, "Jekyll and Hyde Revisited: Paradoxes in the Appreciation of Drug Experiences and Their Effects on Creativity," *Journal of Psychoactive Drugs* 34, no. 3 (September 2002): 249–262, https://doi.org/10.1080/02791072.2002.10399961;

Jos ten Berge, "Breakdown or Breakthrough? A History of European Research into Drugs and Creativity," *Journal of Creative Behavior* 33, no. 4 (1999): 257–276, https://doi.org/10.1002/j.2162-6057.1999.tb 01406.x. For a contemporary quantitative examination using largely self-report measures, see Danielle E. Humphrey et al., "Self-Reported Drug Use and Creativity: (Re)Establishing Layperson Myths," *Imagination, Cognition and Personality* 34, no. 2 (October 2014): 179–201, https://doi.org/10.2190/IC.34.2.f.

25. This is as per the report from the Substance Abuse and Mental Health Services Administration (SAMHSA), Donna M. Bush and Rachel N. Lipari, *Substance Use and Substance Use Disorder by Industry*, The CBHSQ Report (National Survey on Drug Use and Health, 2015), https://www.samhsa.gov/data/sites/default/files/report_1959/Short Report-1959.html.

26. B. Sessa, "Is It Time to Revisit the Role of Psychedelic Drugs in Enhancing Human Creativity?," *Journal of Psychopharmacology* (Oxford) 22, no. 8 (November 2008): 821–827, https://doi.org/10.1177 /0269881108091597.

27. Relevant works that explore the impact of drugs on the artist's overall behavior (not only their creativity) include Fruzsina Iszaj et al., "A Qualitative Study on the Effects of Psychoactive Substance Use upon Artistic Creativity," *Substance Use & Misuse* 53, no. 8 (July 3, 2018): 1275–1280, https://doi.org/10.1080/10826084.2017.1404103; and Iszáj, Griffiths, and Demetrovics, "Creativity and Psychoactive Substance Use." For reports on the reasons for using such drugs, see NIDA, "Hallucinogens DrugFacts."

28. Studies that show higher incidence of substance abuse (drugs generally, not only hallucinogenic substances) among artists have noted in particular that the incidence is markedly high in select creative groups (e.g., writers). They also single out alcohol consumption as a factor of significance: A. M. Ludwig, "Creative Achievement and Psychopathology: Comparison among Professions," *American Journal of Psychotherapy* 46, no. 3 (July 1992): 330–356; Post, "Verbal Creativity, Depression and Alcoholism"; Kyaga et al., "Mental Illness, Suicide and Creativity."

29. A fascinating range of first-person excerpts are showcased in this valuable book chapter: S. Krippner, "Psychedelics, Hypnosis, and Creativity," in *Altered States of Consciousness*, ed. Charles T. Tart, 3rd ed. (New York: HarperCollins, 1990), 324–349; as well in the books penned by early pioneers in LSD research: Fadiman, *The Psychedelic Explorer's Guide*; Marlene Dobkin de Rios and Oscar Janiger, *LSD, Spirituality, and the Creative Process* (Rochester, VT: Park Street Press, 2003).

30. A good overview of prevalence data from national and international drug surveys: Matthew W. Johnson et al., "Classic Psychedelics: An Integrative Review of Epidemiology, Therapeutics, Mystical Experience, and Brain Network Function," *Pharmacology & Therapeutics* 197 (May 2019): 83–102, https://doi.org/10.1016/j.pharmthera.2018.11.010.

31. As a case in point, these studies were conducted at a psilocybin retreat and a microdosing event run by the Dutch Psychedelic Society, respectively: Natasha L. Mason et al., "Sub-Acute Effects of Psilocybin on Empathy, Creative Thinking, and Subjective Well-Being," *Journal of Psychoactive Drugs* 51, no. 2 (June 2019): 123–134, https://doi.org/10.1080/02791072.2019.1580804; Luisa Prochazkova et al., "Exploring the Effect of Microdosing Psychedelics on Creativity in an Open-Label Natural Setting," *Psychopharmacology* 235, no. 12 (December 2018): 3401–3413, https://doi.org/10.1007/s00213-018-5049-7.

32. An exploration of personality characteristics of the people who volunteer for clinical trials indicates lower levels of anxiety and higher levels of openness and impulsiveness: L. Almeida et al., "Personality Characteristics of Volunteers in Phase 1 Studies and Likelihood of Reporting Adverse Events," *International Journal of Clinical Pharmacology and Therapeutics* 46, no. 7 (2008): 340–348, https://doi.org/10.5414/cpp46340; Luis Almeida et al., "Who Volunteers for Phase I Clinical Trials? Influences of Anxiety, Social Anxiety and Depressive Symptoms on Self-Selection and the Reporting of Adverse Events," *European Journal of Clinical Pharmacology* 64, no. 6 (2008): 575–582, https://doi.org/10.1007/s00228-008-0468-8.

33. Robert R. McCrae, "Creativity, Divergent Thinking, and Openness to Experience," *Journal of Personality and Social Psychology* 52, no.

6 (1987): 1258–1265, https://doi.org/10.1037/0022-3514.52.6.1258; Scott Barry Kaufman et al., "Openness to Experience and Intellect Differentially Predict Creative Achievement in the Arts and Sciences," *Journal of Personality* 84, no. 2 (April 2016): 248–258, https://doi.org /10.1111/jopy.12156.

34. In sober cannabis users, enhanced performance on convergent creative tasks was explained by differences in openness to experience: Emily M. LaFrance and Carrie Cuttler, "Inspired by Mary Jane? Mechanisms Underlying Enhanced Creativity in Cannabis Users," *Consciousness and Cognition* 56 (November 2017): 68–76, https://doi .org/10.1016/j.concog.2017.10.009. Openness to experience strongly predicted creativity across all scales and much more so than any measured facet of drug use: Humphrey et al., "Self-Reported Drug Use and Creativity."

35. An examination from neuroimaging research on ecstasy where the average consumption level of pills reported in study participants was 720 percent higher than those reported in global drug surveys: Balázs Szigeti et al., "Are Ecstasy Induced Serotonergic Alterations Overestimated for the Majority of Users?," *Journal of Psychopharmacology (Oxford)* 32, no. 7 (July 2018): 741–748, https://doi.org/10.1177 /0269881118767646.

36. An important early study that showed significant personality differences in people who chose to take or not take LSD in a study: J. H. Bottrill, "Personality Change in LSD Users," *Journal of General Psychology* 80, no. 2 (April 1969): 157–161, https://doi.org/10.1080 /00221309.1969.9710481.

37. Among early scholars, even the most avid proponents of the beneficial consequences of the psychedelic experience were cautious and measured in their claims. See Fadiman, *The Psychedelic Explorer's Guide*; Dobkin de Rios and Janiger, *LSD, Spirituality, and the Creative Process*.

38. Franz X. Vollenweider and Katrin H. Preller, "Psychedelic Drugs: Neurobiology and Potential for Treatment of Psychiatric Disorders," *Nature Reviews Neuroscience* 21, no. 11 (November 2020): 611–624, https://doi.org/10.1038/s41583-020-0367-2; Lionel Barnett et al.,

"Decreased Directed Functional Connectivity in the Psychedelic State," *NeuroImage* 209 (April 2020): 116462, https://doi.org/10.1016/j.neuroimage.2019.116462; Felix Müller et al., "Chapter 6—Advances and Challenges in Neuroimaging Studies on the Effects of Serotonergic Hallucinogens: Contributions of the Resting Brain," in *Psychedelic Neuroscience*, ed. Tanya Calvey, Progress in Brain Research, vol. 242 (Elsevier, 2018), 159–177, https://doi.org/10.1016/bs.pbr.2018.08.004.

39. For an overview on precuneus functions particularly in relation to consciousness, see Andrea E. Cavanna, "The Precuneus and Consciousness," *CNS Spectrums* 12, no. 7 (July 2007): 545–552, https://doi.org/10.1017/s1092852900021295. For a recent overview of thalamus structure and function, see Tyler J. Torrico and Sunil Munakomi, "Neuroanatomy, Thalamus," in *StatPearls* (Treasure Island, FL: StatPearls Publishing, 2021), http://www.ncbi.nlm.nih.gov/books/NBK 542184/.

40. Robin L. Carhart-Harris et al., "Neural Correlates of the LSD Experience Revealed by Multimodal Neuroimaging," *Proceedings of the National Academy of Sciences* 113, no. 17 (2016): 4853–4858, https://doi.org/10.1073/pnas.1518377113.

41. A selection of recent empirical work that makes a case for the role of these regions in spatial/contextual/scene-based memory processes: Sam C. Berens, Bárður H. Joensen, and Aidan J. Horner, "Tracking the Emergence of Location-based Spatial Representations in Human Scene-Selective Cortex," *Journal of Cognitive Neuroscience* 33, no. 3 (March 2021): 445–462, https://doi.org/10.1162/jocn_a_01654; Andrej Bicanski and Neil Burgess, "Neuronal Vector Coding in Spatial Cognition," *Nature Reviews Neuroscience* 21, no. 9 (September 2020): 453–470, https://doi.org/10.1038/s41583-020-0336-9; Russell A. Epstein et al., "The Cognitive Map in Humans: Spatial Navigation and Beyond," *Nature Neuroscience* 20, no. 11 (October 26, 2017): 1504–1513, https://doi.org/10.1038/nn.4656; David J. Bucci and Siobhan Robinson, "Toward a Conceptualization of Retrohippocampal Contributions to Learning and Memory," *Neurobiology of Learning and Memory* 116 (December 2014): 197–207, https://doi.org/10.1016/j.nlm.2014.05.007.

42. Sara Dell'Erba, David J. Brown, and Michael J. Proulx, "Synesthetic Hallucinations Induced by Psychedelic Drugs in a Congenitally Blind Man," *Consciousness and Cognition* 60 (April 2018): 127–132, https://doi.org/10.1016/j.concog.2018.02.008.

43. Andrew R. Gallimore, "Restructuring Consciousness—the Psychedelic State in Light of Integrated Information Theory," *Frontiers in Human Neuroscience* 9 (2015): 346, https://doi.org/10.3389/fnhum.2015.00346.

44. Andrea Alamia et al., "DMT Alters Cortical Travelling Waves," *eLife* 9 (October 12, 2020): e59784, https://doi.org/10.7554/eLife.59784.

45. Calvin Ly et al., "Psychedelics Promote Structural and Functional Neural Plasticity," *Cell Reports* 23, no. 11 (June 12, 2018): 3170–3182, https://doi.org/10.1016/j.celrep.2018.05.022.

46. David A. Martin and Charles D. Nichols, "Psychedelics Recruit Multiple Cellular Types and Produce Complex Transcriptional Responses within the Brain," *EBioMedicine* 11 (September 2016): 262–277, https://doi.org/10.1016/j.ebiom.2016.08.049.

47. Barron, *Creativity and Psychological Health*, 256–257.

48. Iszáj, Griffiths, and Demetrovics, "Creativity and Psychoactive Substance Use."

49. This paper reports the findings of two studies in which participants who were presented alcohol cues (study 1) and marijuana cues (study 2) demonstrated better performance on a divergent thinking task (on measures of ideational fluency and ideational flexibility) if they expected these psychoactive substances to boost creativity: Joshua A. Hicks et al., "Expecting Innovation: Psychoactive Drug Primes and the Generation of Creative Solutions," *Experimental and Clinical Psychopharmacology* 19, no. 4 (August 2011): 314–320, https://doi.org/10.1037/a0022954.

50. Jay A. Olson et al., "Tripping on Nothing: Placebo Psychedelics and Contextual Factors," *Psychopharmacology* 237, no. 5 (May 2020): 1371–1382, https://doi.org/10.1007/s00213-020-05464-5.

51. Oscar Janiger and Marlene Dobkin de Rios, "LSD and Creativity," *Journal of Psychoactive Drugs* 21, no. 1 (March 1989): 129–134, https://doi.org/10.1080/02791072.1989.10472150.

52. The terms "creative" and "imaginative" are often used interchangeably in general parlance. In scientific research, these concepts are distinguished from one another. The element of novelty or originality is central to deeming something creative. Heavily drawing on sensory imagery or fantasy is what is associated with imaginativeness in this context.

53. "The pictures looked strange, but not because they bore the influence of the drugs, but because a seriously ill person can only paint with difficulty." Hunderwasser in Hartmann, *Malerei aus Bereichen des Unbewussten*, 229–230 (author's translation).

54. Manfred Spitzer et al., "Increased Activation of Indirect Semantic Associations under Psilocybin," *Biological Psychiatry* 39, no. 12 (June 15, 1996): 1055–1057, https://doi.org/10.1016/0006-3223(95)00418-1; Neiloufar Family et al., "Semantic Activation in LSD: Evidence from Picture Naming," *Language, Cognition and Neuroscience* 31, no. 10 (2016): 1320–1327, https://doi.org/10.1080/23273798.2016.1217030.

55. Katrin H. Preller et al., "The Fabric of Meaning and Subjective Effects in LSD-Induced States Depend on Serotonin 2A Receptor Activation," *Current Biology* 27, no. 3 (2017): 451–457, https://doi.org/10.1016/j.cub.2016.12.030.

56. Fadiman, *The Psychedelic Explorer's Guide*; Dobkin de Rios and Janiger, *LSD, Spirituality, and the Creative Process*.

57. Krippner, "Psychedelics, Hypnosis, and Creativity."

58. Schmid and Liechti, "Long-Lasting Subjective Effects of LSD in Normal Subjects."

59. Willis W. Harman et al., "Psychedelic Agents in Creative Problem-Solving: A Pilot Study," *Psychological Reports* 19, no. 1 (August 1966): 211–227, https://doi.org/10.2466/pr0.1966.19.1.211.

60. A reanalysis of this data within this review has since called these findings into question: Matthew J. Baggott, "Psychedelics and

Creativity: A Review of the Quantitative Literature" (PeerJ Inc., June 30, 2015), https://doi.org/10.7287/peerj.preprints.1202v1.

61. N. L. Mason et al., "Spontaneous and Deliberate Creative Cognition during and after Psilocybin Exposure," *Translational Psychiatry* 11, no. 1 (April 8, 2021): 209, https://doi.org/10.1038/s41398-021-01335-5.

62. I-creativity or individual creativity (formerly referred to as psychological/P-creativity) or the subjective experience of generating an idea that is novel, relevant, and surprising to oneself is typically distinguished from historical/H-creativity. In the latter case, an idea is experienced as creative not only by the person generating the idea, but also en masse by the recipients who are privy to the idea. For more, see Boden, "Creativity and Biology."

63. George Wickes and Ray Frazer, "The Art of Fiction XXIV: Aldous Huxley," *Paris Review*, no. 23 (Spring 1960): 56–80.

64. David Nutt, David Erritzoe, and Robin Carhart-Harris, "Psychedelic Psychiatry's Brave New World," *Cell* 181, no. 1 (April 2, 2020): 24–28, https://doi.org/10.1016/j.cell.2020.03.020; Kuypers et al., "Microdosing Psychedelics."

65. David B. Yaden and Roland R. Griffiths, "The Subjective Effects of Psychedelics Are Necessary for Their Enduring Therapeutic Effects," *ACS Pharmacology & Translational Science* 4, no. 2 (December 10, 2020): 568–572, https://doi.org/10.1021/acsptsci.0c00194; Alan K. Davis, Frederick S. Barrett, and Roland R. Griffiths, "Psychological Flexibility Mediates the Relations between Acute Psychedelic Effects and Subjective Decreases in Depression and Anxiety," *Journal of Contextual Behavioral Science* 15 (January 2020): 39–45, https://doi.org/10.1016/j.jcbs.2019.11.004.

66. Betty Eisner, "Set, Setting, and Matrix," *Journal of Psychoactive Drugs* 29, no. 2 (June 1997): 213–216, https://doi.org/10.1080/02791072.1997.10400190; S. A. McWilliams and R. J. Tuttle, "Long-Term Psychological Effects of LSD," *Psychological Bulletin* 79, no. 6 (June 1973): 341–351, https://doi.org/10.1037/h0034411.

67. This functional neuroimaging study evidences the importance of the salience network in creative ideation: Abraham et al., "Creative Conceptual Expansion."

68. The case for cautious optimism is outlined succinctly in David B. Yaden, Mary E. Yaden, and Roland R. Griffiths, "Psychedelics in Psychiatry—Keeping the Renaissance from Going off the Rails," *JAMA Psychiatry* 78, no. 5 (2021): 469, https://doi.org/10.1001/jama psychiatry.2020.3672; and L.-S. Camilla d'Angelo, George Savulich, and Barbara J. Sahakian, "Lifestyle Use of Drugs by Healthy People for Enhancing Cognition, Creativity, Motivation and Pleasure," *British Journal of Pharmacology* 174, no. 19 (October 2017): 3257–3267, https://doi.org/10.1111/bph.13813.

69. Hofmann, *LSD, My Problem Child*.

Chapter 4

1. Oliver Sacks, *An Anthropologist on Mars: Seven Paradoxical Tales* (New York: Knopf, 1995), xvi.

2. Lev Semenovich Vygotsky, *The Collected Works of L. S. Vygotsky, Vol. 2: The Fundamentals of Defectology (Abnormal Psychology and Learning Disabilities)*, ed. Robert W. Rieber and Aaron S. Carton (New York: Plenum Press, 1993), 34.

3. In fact, Hughlings Jackson noted the following in the late 1800s: "The symptomatology of nervous diseases is a double condition; there is a negative and there is a positive element in every case. Evolution not being entirely reversed, some level of evolution is left." J. Hughlings Jackson, "The Croonian Lectures on Evolution and Dissolution of the Nervous System," *British Medical Journal* 1, no. 1213 (1884): 591.

4. Vygotsky, *Vol. 2: Fundamentals of Defectology*, 30, 31, 34; italics in original.

5. Alexander Luria (1902–1977), a great proponent of what he termed "romantic science," was an early advocate of the power of

brain plasticity. He explained the difference between the classical approach and his adopted approach as follows: "'Romantic scholars' traits, attitudes, and strategies are just the opposite. They do not follow the path of reductionism, which is the leading philosophy of the classical group. Romantics in science want neither to split living reality into its elementary components nor to represent the wealth of life's concrete events in abstract models that lose the properties of the phenomena themselves. It is of the utmost importance to romantics to preserve the wealth of living reality, and they aspire to a science that retains this richness. Of course, romantic scholars and romantic science have their shortcomings. Romantic science typically lacks the logic and does not follow the careful, consecutive, step-by-step reasoning that is characteristic of classical science, nor does it easily reach firm formulations and universally applicable laws. Sometimes logical step-by-step analysis escapes romantic scholars, and on occasion, they let artistic preferences and intuitions take over. Frequently their descriptions not only precede explanation but replace it. I have long puzzled over which of the two approaches, in principle, leads to a better understanding of living reality." See A. R. Luria, Sheila Cole, and Michael Cole, *The Making of Mind: A Personal Account of Soviet Psychology* (Cambridge, MA: Harvard University Press, 1979).

6. Blind individuals are also better able to localize sounds monoaurally than sighted individuals: N. Lessard et al., "Early-Blind Human Subjects Localize Sound Sources Better than Sighted Subjects," *Nature* 395 (September 17, 1998): 278–280, https://doi.org/10.1038/26228.

7. Daniel Goldreich and Ingrid M. Kanics, "Tactile Acuity Is Enhanced in Blindness," *Journal of Neuroscience* 23, no. 8 (April 15, 2003): 3439–3445.

8. Blind individuals not only demonstrated better free identification of odors, but also enhanced odor discrimination and categorization: Isabel Cuevas et al., "Odour Discrimination and Identification Are Improved in Early Blindness," *Neuropsychologia* 47, no. 14 (December 2009): 3079–3083, https://doi.org/10.1016/j.neuropsychologia.2009.07.004.

9. Annie H. Takeuchi and Stewart H. Hulse, "Absolute Pitch.," *Psychological Bulletin* 113, no. 2 (1993): 345–361, https://doi.org/10.1037/0033-2909.113.2.345.

10. Maria Dimatati et al., "Exploring the Impact of Congenital Visual Impairment on the Development of Absolute Pitch Using a New Online Assessment Tool: A Preliminary Study," *Psychomusicology: Music, Mind, and Brain* 22, no. 2 (December 2012): 129–133, https://doi.org/10.1037/a0030857. For media coverage in relation to these findings, see Lucy Tobin, "Why Musical Talent Can Stem from Visual Impairment," *Guardian*, May 18, 2010, sec. Education, https://www.theguardian.com/education/2010/may/18/musical-talent-link-with-blindness.

11. Out of 612 musicians, 92 self-reported perfect pitch: S. Baharloo et al., "Absolute Pitch: An Approach for Identification of Genetic and Nongenetic Components," *American Journal of Human Genetics* 62, no. 2 (February 1998): 224–231, https://doi.org/10.1086/301704.

12. Roy H. Hamilton, Alvaro Pascual-Leone, and Gottfried Schlaug, "Absolute Pitch in Blind Musicians," *Neuroreport* 15, no. 5 (April 9, 2004): 803–806, https://doi.org/10.1097/00001756-200404090-00012.

13. N. Kapur, "Paradoxical Functional Facilitation in Brain–Behaviour Research: A Critical Review," *Brain: A Journal of Neurology* 119, no. 5 (October 1996): 1775, https://doi.org/10.1093/brain/119.5.1775.

14. For a comprehensive overview, see Narinder Kapur, ed., *The Paradoxical Brain* (Cambridge: Cambridge University Press, 2011).

15. Moheb Costandi, *Neuroplasticity* (Cambridge, MA: MIT Press, 2016), 2.

16. The others include hierarchical representation, network organization, fine-grained temporal synchrony, multifunctionality, degeneracy (i.e., varied routes to function), homeostatis and stability, competitive dynamics, nonlinearity, and itinerant dynamics: Narinder Kapur et al., "The Paradoxical Brain—So What?," in *The Paradoxical Brain*, ed. Narinder Kapur (Cambridge: Cambridge University Press, 2011), 418–434, https://doi.org/10.1017/CBO9780511978098.026.

17. Nadine Gaab et al., "Neural Correlates of Absolute Pitch Differ between Blind and Sighted Musicians," *Neuroreport* 17, no. 18 (December 18, 2006): 1853–1857, https://doi.org/10.1097/WNR.0b013e328 0107bee.

18. Amir Amedi et al., "The Occipital Cortex in the Blind: Lessons about Plasticity and Vision," *Current Directions in Psychological Science* 14, no. 6 (2005): 306–311.

19. For an overview on "romantic" views of science, including criticisms proponents have faced, see Martin Halliwell, *Romantic Science and the Experience of Self: Transatlantic Crosscurrents from William James to Oliver Sacks*, Studies in European Cultural Transition, vol. 2 (Aldershot, Hants; Brookfield, VT: Ashgate, 1999).

20. Darold A. Treffert, "The Savant Syndrome: An Extraordinary Condition. A Synopsis: Past, Present, Future," *Philosophical Transactions of the Royal Society B* 364, no. 1522 (May 27, 2009): 1351, https://doi.org /10.1098/rstb.2008.0326.

21. Timo Saloviita, Liisa Ruusila, and Unto Ruusila, "Incidence of Savant Syndrome in Finland," *Perceptual and Motor Skills* 91, no. 1 (August 2000): 120–122, https://doi.org/10.2466/pms.2000.91.1.120.

22. The more conservative estimate is espoused in Beate Hermelin, *Bright Splinters of the Mind: A Personal Story of Research with Autistic Savants* (London; Philadelphia: J. Kingsley, 2001).

23. For overviews of savant syndrome that are both comprehensive and evocative, see Darold A. Treffert, *Extraordinary People: Understanding "Idiot Savants"* (New York: Harper & Row, 1989); Darold A. Treffert, *Islands of Genius: The Bountiful Mind of the Autistic, Acquired, and Sudden Savant* (London: Jessica Kingsley Publishers, 2010).

24. Leon K. Miller, "The Savant Syndrome: Intellectual Impairment and Exceptional Skill.," *Psychological Bulletin* 125, no. 1 (1999): 37, https://doi.org/10.1037/0033-2909.125.1.31.

25. This is specifically proposed for autistic savants in Laurent Mottron, Michelle Dawson, and Isabelle Soulières, "Enhanced Perception

in Savant Syndrome: Patterns, Structure and Creativity," *Philosophical Transactions of the Royal Society B* 364, no. 1522 (May 27, 2009): 1385–1391, https://doi.org/10.1098/rstb.2008.0333.

26. M. J. Howe, J. W. Davidson, and J. A. Sloboda, "Innate Talents: Reality or Myth?," *Behavioral and Brain Sciences* 21, no. 3 (June 1998): 399–407; discussion 407–442, https://doi.org/10.1017/s0140525x98 00123x.

27. Neil O'Connor and Beate Hermelin, "Talents and Preoccupations in Idiots-Savants," *Psychological Medicine* 21, no. 4 (November 1991): 959–964, https://doi.org/10.1017/S0033291700029949.

28. For a general overview, see Leon K. Miller, "What the Savant Syndrome Can Tell Us about the Nature and Nurture of Talent," *Journal for the Education of the Gifted* 28, no. 3–4 (March 2005): 361–373, https://doi.org/10.4219/jeg-2005-340.

29. Scott Feinberg, "The World's Busiest Oscar Has Traveled Over 3 Million Miles," *Hollywood Reporter*, February 21, 2017, https://www .hollywoodreporter.com/movies/movie-news/worlds-busiest-oscar -has-traveled-3-million-miles-977086/.

30. "Kim Peek; American Savant with a Photographic Memory Who Was the Inspiration for the Film Rain Man and Was Rated a Genius in 15 Subjects," *The Times* (London), December 23, 2009.

31. Darold A. Treffert and Daniel D. Christensen, "Inside the Mind of a Savant," *Scientific American* 293, no. 6 (2005): 108–113.

32. Famous cases in point of prodigious savants, all with the accompanying developmental disability of autism, who undoubtedly exhibit creativity include Stephen Wiltshire (http://www.stephen wiltshire.co.uk/), Matt Savage (https://www.savagerecords.com/word pressnew/), and Daniel Tammet (http://www.danieltammet.net/index .php).

33. Treffert, *Islands of Genius*, 125.

34. Allan Snyder, Terry Bossomaier, and D. John Mitchell, "Concept Formation: 'Object' Attributes Dynamically Inhibited from Conscious

Awareness," *Journal of Integrative Neuroscience* 3, no. 1 (March 2004): 31–46, https://doi.org/10.1142/s0219635204000361.

35. Allan Snyder, "Explaining and Inducing Savant Skills: Privileged Access to Lower Level, Less-Processed Information," *Philosophical Transactions of the Royal Society B* 364, no. 1522 (May 27, 2009): 1399–1405, https://doi.org/10.1098/rstb.2008.0290.

36. Lucie Bouvet et al., "Veridical Mapping in Savant Abilities, Absolute Pitch, and Synesthesia: An Autism Case Study," *Frontiers in Psychology* 5 (February 18, 2014): 1, https://doi.org/10.3389/fpsyg.2014.00106; Laurent Mottron et al., "Veridical Mapping in the Development of Exceptional Autistic Abilities," *Neuroscience & Biobehavioral Reviews* 37, no. 2 (February 2013): 210, https://doi.org/10.1016/j.neubiorev.2012.11.016.

37. Darold A. Treffert, "Accidental Genius," *Scientific American* 311, no. 2 (2014): 52–57.

38. Tara Boyle et al., "Stroke of Genius: How Derek Amato Became a Musical Savant," *NPR*, February 23, 2016, https://www.npr.org/2016/02/22/467680296/stroke-of-genius-how-derek-amato-became-a-musical-savant.

39. Jason Padgett and Maureen Ann Seaberg, *Struck by Genius: How a Brain Injury Made Me a Mathematical Marvel* (Boston: Mariner Books, 2014).

40. A registry of more than three hundred savants pegged the incidence rate of acquired savantism at 10 percent: Darold A. Treffert and David L. Rebedew, "The Savant Syndrome Registry: A Preliminary Report," *WMJ: Official Publication of the State Medical Society of Wisconsin* 114, no. 4 (August 2015): 158–162.

41. Thomas A. Pollak, Catherine M. Mulvenna, and Mark F. Lythgoe, "De Novo Artistic Behaviour Following Brain Injury," *Neurological Disorders in Famous Artists—Part 2* 22 (2007): 75–88, https://doi.org/10.1159/000102873.

42. Matthew A. Lambon Ralph et al., "The Neural and Computational Bases of Semantic Cognition," *Nature Reviews Neuroscience* 18, no. 1 (January 2017): 42–55, https://doi.org/10.1038/nrn.2016.150.

43. In the literature svPPA is often referred to as temporal lobe variant-FTD: Zachary A. Miller and Bruce L. Miller, "Chapter 5—Artistic Creativity and Dementia," in *The Fine Arts, Neurology, and Neuroscience,* ed. Stanley Finger et al., Progress in Brain Research, vol. 204 (Elsevier, 2013): 99–112, https://doi.org/10.1016/B978-0-444-63287-6.00005-1.

44. The focus in this chapter is only limited to this subtype as findings in relation to other FTD subtypes and other dementias have been mixed at best: B. L. Miller et al., "Emergence of Artistic Talent in Frontotemporal Dementia," *Neurology* 51, no. 4 (October 1998): 978–982, https://doi.org/10.1212/WNL.51.4.978; B. L. Miller et al., "Enhanced Artistic Creativity with Temporal Lobe Degeneration," *Lancet* 348, no. 9043 (December 21, 1996): 1744–1745.

45. Anjan Chatterjee, "The Neuropsychology of Visual Artistic Production," *Neuropsychologia* 42, no. 11 (2004): 1568–1583, https://doi.org/10.1016/j.neuropsychologia.2004.03.011.

46. Patient 4 and 5 in B. L. Miller et al., "Functional Correlates of Musical and Visual Ability in Frontotemporal Dementia," *British Journal of Psychiatry* 176 (May 2000): 458–463.

47. Teresa Q. Wu et al., "Verbal Creativity in Semantic Variant Primary Progressive Aphasia," *Neurocase* 21, no. 1 (February 2015): 73–78, https://doi.org/10.1080/13554794.2013.860179.

48. See also note 81. Given that the emergence of de novo artistic skills occurs in the context of language dysfunction, the author proposes that this behavior reflects the drive to express oneself and communicate: Dahlia W. Zaidel, "Creativity, Brain, and Art: Biological and Neurological Considerations," *Frontiers in Human Neuroscience* 8 (2014), https://doi.org/10.3389/fnhum.2014.00389. This sentiment is shared by others—"creativity would at least be an expression of this person, a sufferer, a communication of the inner life": Felix Geser et al., "Emergent Creativity in Frontotemporal Dementia," *Journal of Neural Transmission* 128, no. 3 (March 2021): 290, https://doi.org/10.1007/s00702-021-02325-z.

49. Anita Thapar, Miriam Cooper, and Michael Rutter, "Neurodevelopmental Disorders," *The Lancet Psychiatry* 4, no. 4 (April 2017): 339–346, https://doi.org/10.1016/S2215-0366(16)30376-5.

50. Miriam D. Lense et al., "Rhythm and Timing as Vulnerabilities in Neurodevelopmental Disorders," *Philosophical Transactions of the Royal Society B* 376, no. 1835 (October 11, 2021): 20200327, https://doi.org/10.1098/rstb.2020.0327.

51. For an example of a study of this finding from dyslexia research, see Catya von Károlyi et al., "Dyslexia Linked to Talent: Global Visual-Spatial Ability," *Brain and Language* 85, no. 3 (June 2003): 427–431, https://doi.org/10.1016/s0093-934x(03)00052-x. For more information on this positive bias in information processing for autism, see Steven Dakin and Uta Frith, "Vagaries of Visual Perception in Autism," *Neuron* 48, no. 3 (November 3, 2005): 497–507, https://doi.org/10.1016/j.neuron.2005.10.018.

52. Richard L. Masland, "The Advantages of Being Dyslexic," *Bulletin of the Orton Society* 26 (1976): 10–18.

53. Art students enrolled in university showed more signs of dyslexia than non-art students: Ulrika Wolff and Ingvar Lundberg, "The Prevalence of Dyslexia among Art Students," *Dyslexia* (Chichester, England) 8, no. 1 (March 2002): 34–42, https://doi.org/10.1002/dys.211. While it is tempting to conclude that this is evidence showing that dyslexia is associated with a heightened interest in taking up the visual arts as a profession, the grounds for the choice of subject have not been systematically investigated in quantitative studies. That other factors need to be considered was suggested by a qualitative analysis of thirteen art students with dyslexia that revealed that while some of the students chose to study art in higher education because they enjoyed art making, others did so because they felt had few options apart from art to pursue: Alison M. Bacon and Samantha Bennett, "Dyslexia in Higher Education: The Decision to Study Art," *European Journal of Special Needs Education* 28, no. 1 (February 2013): 19–32, https://doi.org/10.1080/08856257.2012.742748.

54. Rebecca Chamberlain et al., "Meta-analytic Findings Reveal Lower Means but Higher Variances in Visuospatial Ability in Dyslexia," *British Journal of Psychology* 109, no. 4 (November 2018): 897–916, https://doi.org/10.1111/bjop.12321.

55. This meta-analysis examined creative test performance in children and adults separately. Out of the six studies on creativity and dyslexia in adulthood, five showed better performance in dyslexic groups: Nadyanna M. Majeed, Andree Hartanto, and Jacinth J. X. Tan, "Developmental Dyslexia and Creativity: A Meta-analysis," *Dyslexia* 27, no. 2 (May 2021): 187–203, https://doi.org/10.1002/dys.1677; this pattern was confirmed in Florina Erbeli, Peng Peng, and Marianne Rice, "No Evidence of Creative Benefit Accompanying Dyslexia: A Meta-analysis," *Journal of Learning Disabilities* 55, no. 3 (April 24, 2021), https://doi.org/10.1177/00222194211010350.

56. Patricia Howlin et al., "Savant Skills in Autism: Psychometric Approaches and Parental Reports," *Philosophical Transactions of the Royal Society B* 364, no. 522 (May 27, 2009): 1364, https://doi.org/10.1098/rstb.2008.0328.

57. Not everyone agrees. Several scholars have posited that autism is marked by impoverished reality-centered imagination skills: J. Craig and S. Baron-Cohen, "Creativity and Imagination in Autism and Asperger Syndrome," *Journal of Autism and Developmental Disorders* 29, no. 4 (August 1999): 319–326, https://doi.org/10.1023/a:1022163403479; Bernard Crespi et al., "Imagination in Human Social Cognition, Autism, and Psychotic-Affective Conditions," *Cognition* 150 (May 2016): 181–199, https://doi.org/10.1016/j.cognition.2016.02.001. However, others have remarked that double standards are often associated with such views given that realistic styles are only viewed as a negative instantiation of imaginative capacities when the drawings are made by participants with autism, but not when made by neurotypical people: Ilona Roth, "Autism, Creativity and Aesthetics," *Qualitative Research in Psychology* 17, no. 4 (October 2020): 498–508, https://doi.org/10.1080/14780887.2018.1442763; Gillian J. Furniss, "Celebrating the Artmaking of Children with Autism," *Art Education* 61, no. 5 (September 2008): 8–12, https://doi.org/10.1080/00043125.2008.11518990.

58. Pamela Heaton, Beate Hermelin, and Linda Pring, "Autism and Pitch Processing: A Precursor for Savant Musical Ability?," *Music Perception: An Interdisciplinary Journal* 15, no. 3 (1998): 291–305, https://doi.org/10.2307/40285769. In fact, superior pitch processing, particularly pitch discrimination and categorization, are enhanced in

nonsavant individuals with autism: Pamela Heaton, "Assessing Musical Skills in Autistic Children Who Are Not Savants," *Philosophical Transactions of the Royal Society B* 364, no. 1522 (2009): 1443–1447.

59. Walter A. Brown et al., "Autism-Related Language, Personality, and Cognition in People with Absolute Pitch: Results of a Preliminary Study," *Journal of Autism and Developmental Disorders* 33, no. 2 (April 2003): 163–167; discussion 169, https://doi.org/10.1023/a:1022987309913.

60. T. Wenhart et al., "Autistic Traits, Resting-State Connectivity, and Absolute Pitch in Professional Musicians: Shared and Distinct Neural Features," *Molecular Autism* 10, no. 1 (May 2, 2019): 20, https://doi.org/10.1186/s13229-019-0272-6.

61. This advantage did not extend to visual tasks: Paola Pennisi et al., "Autism, Autistic Traits and Creativity: A Systematic Review and Meta-analysis," *Cognitive Processing* 22, no. 1 (2021): 1–36, https://doi.org/10.1007/s10339-020-00992-6.

62. On the other hand, there was some evidence of familial associations as siblings of people autism were overrepresented in creative professions: Kyaga et al., "Mental Illness, Suicide and Creativity."

63. Kendra S. Knudsen, Susan Y. Bookheimer, and Robert M. Bilder, "Is Psychopathology Elevated in Big-C Visual Artists and Scientists?," *Journal of Abnormal Psychology* 128, no. 4 (May 2019): 273–283, https://doi.org/10.1037/abn0000416.

64. Dakin and Frith, "Vagaries of Visual Perception in Autism"; David R. Simmons et al., "Vision in Autism Spectrum Disorders," *Vision Research* 49, no. 22 (November 2009): 2705–2739, https://doi.org/10.1016/j.visres.2009.08.005; Zsuzsa Kaldy et al., "The Mechanisms Underlying the ASD Advantage in Visual Search," *Journal of Autism and Developmental Disorders* 46, no. 5 (May 2016): 1513–1527, https://doi.org/10.1007/s10803-013-1957-x.

65. Kyaga et al., "Mental Illness, Suicide and Creativity."

66. Martine Hoogman et al., "Creativity and ADHD: A Review of Behavioral Studies, the Effect of Psychostimulants and Neural

Underpinnings," *Neuroscience & Biobehavioral Reviews* 119 (December 2020): 66–85, https://doi.org/10.1016/j.neubiorev.2020.09.029.

67. Michelle A Pievsky and Robert E McGrath, "The Neurocognitive Profile of Attention-Deficit/Hyperactivity Disorder: A Review of Meta-analyses," *Archives of Clinical Neuropsychology* 33, no. 2 (March 2018): 143–157, https://doi.org/10.1093/arclin/acx055; A. Marije Boonstra et al., "Executive Functioning in Adult ADHD: A Meta-analytic Review," *Psychological Medicine* 35, no. 8 (August 2005): 1097–1108, https://doi.org/10.1017/S003329170500499X.

68. Reduced latent disinhibition was found to be associated with higher creative achievement: Shelley Carson, "Latent Inhibition and Creativity," in *Latent Inhibition: Cognition, Neuroscience and Applications to Schizophrenia* (New York: Cambridge University Press, 2010), 183–198, https://doi.org/10.1017/CBO9780511730184.010. Experimentally induced disinhibition improves select aspects of divergent thinking: Rémi Radel et al., "The Role of (Dis)Inhibition in Creativity: Decreased Inhibition Improves Idea Generation," *Cognition* 134 (January 2015): 110–120, https://doi.org/10.1016/j.cognition.2014.09.001.

69. Anna Abraham et al., "Creative Thinking in Adolescents with Attention Deficit Hyperactivity Disorder (ADHD)," *Child Neuropsychology: A Journal on Normal and Abnormal Development in Childhood and Adolescence* 12, no. 2 (2006): 111–123, https://doi.org/10.1080/09297040500320691; Holly A. White, "'Thinking 'Outside the Box': Unconstrained Creative Generation in Adults with Attention Deficit Hyperactivity Disorder," *Journal of Creative Behavior* 54, no. 2 (June 2020): 472–883.

70. Julien Bogousslavsky and François Boller, eds., *Neurological Disorders in Famous Artists—Part 1*, Frontiers of Neurology and Neuroscience, vol. 19 (Basel; New York: Karger, 2005); Julien Bogousslavsky and M. Hennerici, eds., *Neurological Disorders in Famous Artists—Part 2*, Frontiers of Neurology and Neuroscience vol. 22 (Basel; New York: Karger, 2007); Julien Bogousslavsky et al., eds., *Neurological Disorders in Famous Artists—Part 3*, Frontiers of Neurology and Neuroscience, vol. 27 (Basel: Karger, 2010); Julien Bogousslavsky and Laurent Tatu, eds., *Neurological Disorders in Famous Artists—Part 4*, Frontiers of Neurology

and Neuroscience, vol. 43 (Basel; New York: Karger, 2018); Stanley Finger et al., eds., *The Fine Arts, Neurology, and Neuroscience: History and Modern Perspectives: Neuro-Historical Dimensions*, Progress in Brain Research, vol. 204 (Elsevier, 2013).

71. Claire Maingon and Laurent Tatu, "Creative Minds in the Aftermath of the Great War: Four Neurologically Wounded Artists," in Bogousslavsky and Tatu, *Neurological Disorders in Famous Artists—Part 4*, 37–46, https://doi.org/10.1159/000490403.

72. Matthew Pelowski et al., "Can We Really 'Read' Art to See the Changing Brain? A Review and Empirical Assessment of Clinical Case Reports and Published Artworks for Systematic Evidence of Quality and Style Changes Linked to Damage or Neurodegenerative Disease," *Physics of Life Reviews* 43 (December 2022): 32–95, https://doi.org/10.1016/j.plrev.2022.07.005.

73. Alex Forsythe, Tamsin Williams, and Ronan G. Reilly, "What Paint Can Tell Us: A Fractal Analysis of Neurological Changes in Seven Artists.," *Neuropsychology* 31, no. 1 (2017): 1–10, https://doi.org/10.1037/neu0000303; Bartlomiej Piechowski-Jozwiak and Julien Bogousslavsky, "Dementia and Change of Style: Willem de Kooning—Obliteration of Disease Patterns?," in Bogousslavsky and Tatu, *Neurological Disorders in Famous Artists—Part 4*, 164–176, https://doi.org/10.1159/000490447.

74. Maria Pachalska et al., "Rehabilitation of an Artist after Right-Hemisphere Stroke," *Medical Science Monitor: International Medical Journal of Experimental and Clinical Research* 14, no. 10 (October 2008): CS110–124.

75. When comparing the artworks across the trajectory, from ten years before the symptoms appeared (1991) to four years after (2004), the shift in style actually shows a return at the end (*Amsterdam*) to a style exhibited when the artist was at a less experienced stage (*Henning's building*); a return to the former familiar style rather than an evolution to a new style: William W. Seeley et al., "Unravelling Boléro: Progressive Aphasia, Transmodal Creativity and the Right Posterior Neocortex," *Brain: A Journal of Neurology* 131, no. 1 (January 2008): 39–49, https://doi.org/10.1093/brain/awm270.

76. Edwin Ruiz and Patricia Montañés, "Music and the Brain: Gershwin and Shebalin," in J. Bogousslavsky and F. Boller, *Neurological Disorders in Famous Artists—Part 1*, 172–178, https://doi.org/10.1159/isbn.978-3-318-01206-4.

77. Julien Bogousslavsky, "The Last Myth of Giorgio De Chirico: Neurological Art," in Bogousslavsky et al., *Neurological Disorders in Famous Artists—Part 3*, 43, https://doi.org/10.1159/000311190.

78. This is the position put forward in Darold A. Treffert, "Savant Syndrome: Realities, Myths and Misconceptions," *Journal of Autism and Developmental Disorders* 44, no. 3 (2014): 565, https://doi.org/10.1007/s10803-013-1906-8; and in Snyder, "Explaining and Inducing Savant Skills."

79. This is the position is advocated in Zaidel, "Creativity, Brain, and Art," 5.

80. Jon O. Lauring et al., "Why Would Parkinson's Disease Lead to Sudden Changes in Creativity, Motivation, or Style with Visual Art?: A Review of Case Evidence and New Neurobiological, Contextual, and Genetic Hypotheses," *Neuroscience & Biobehavioral Reviews* 100 (May 2019): 129–165, https://doi.org/10.1016/j.neubiorev.2018.12.016.

81. Massimiliano Palmiero, Dina Di Giacomo, and Domenico Passafiume, "Creativity and Dementia: A Review," *Cognitive Processing* 13 (2012): 193–209, https://doi.org/10.1007/s10339-012-0439-y; Cosima Gretton and Dominic H. ffytche, "Art and the Brain: A View from Dementia," *International Journal of Geriatric Psychiatry* 29, no. 2 (February 2014): 111–126, https://doi.org/10.1002/gps.3975.

82. Zaidel, "Creativity, Brain, and Art," 2.

83. *Touched by Genius: A Neurological Look at Creativity*, BBC Worldwide Ltd. (New York: Films Media Group, 2007).

84. "The creative work is a novel work that is accepted as tenable or useful or satisfying by a group in some point in time. . . . By 'novel' I mean that the creative product did not exist previously in precisely the same form. It arises from a reintegration of already existing materials or knowledge, but when it is completed it contains elements

that are new. The extent to which a work is novel depends on the extent to which it deviates from the traditional or the status quo. . . . Often, in studying creativity, we tend to restrict ourselves to a study of the genius because the 'distance' between what he has done and what has existed is quite marked. Such an approach causes us to overlook a necessary distinction between the creative product and the creative experience. The child who fixes the bell on his tricycle for the first time may go through stages that are structurally similar to those which characterize the work of the genius. His finished product, however, is a return to a previously existing state of affairs. The product of an inventor's labor, on the other hand, may strike one as creative immediately because it did not exist previously. In speaking of creativity, therefore, it is necessary to distinguish between internal and external frames of reference": Stein, "Creativity and Culture," 311.

85. I-creativity is popularly referred to as P-creativity (psychological, personal) in line with its original formulation by Margaret Boden in her prior works. See Boden, "Creativity and Biology."

86. Alice Flaherty examines the creative drive, an often ignored variable in the empirical study of creativity: Alice Flaherty, *The Midnight Disease: The Drive to Write, Writer's Block, and the Creative Brain* (Boston: Houghton Mifflin, 2004).

87. Oliver Sacks, "Tourette's Syndrome and Creativity: Exploiting the Ticcy Witticisms and Witty Ticcicisms," *BMJ: British Medical Journal* 305, no. 6868 (1992): 1516.

Chapter 5

1. C. Spearman, *Creative Mind* (London: Nisbet & Co. Ltd., 1930), 14, http://archive.org/details/in.ernet.dli.2015.112562.

2. J. P. Guilford, "Creativity," *American Psychologist* 5, no. 9 (1950): 454, https://doi.org/10.1037/h0063487.

3. Joel N. Shurkin, *Terman's Kids: The Groundbreaking Study of How the Gifted Grow Up* (Boston: Little, Brown, 1992), 190.

4. Albert Camus, "Intelligence and the Scaffold," in *Lyrical and Critical Essays* (New York: Knopf, 1969), 217.

5. For accessible overviews of the many polemic positions on this topic, see Richard J. Haier, *The Neuroscience of Intelligence*, Cambridge Fundamentals of Neuroscience in Psychology (New York: Cambridge University Press, 2017); Stephen Murdoch, *IQ: A Smart History of a Failed Idea* (Hoboken, NJ: J. Wiley and Sons, 2007); Russell T. Warne, *In the Know: Debunking 35 Myths about Human Intelligence* (Cambridge; New York: Cambridge University Press, 2020).

6. For a brief yet expansive overview, see Linda S. Gottfredson, "Pretending That Intelligence Doesn't Matter," *Cerebrum: The Dana Forum on Brain Science* 2, no. 3 (2000): 75–96.

7. Linda S. Gottfredson, "Mainstream Science on Intelligence: An Editorial with 52 Signatories, History and Bibliography," *Intelligence* 24, no. 1 (1997): 13, https://doi.org/10.1016/S0160-2896(97)90011-8.

8. J. E. Drevdahl, "Factors of Importance for Creativity," *Journal of Clinical Psychology* 12, no. 1 (January 1956): 22.

9. Robert Burch, "Charles Sanders Peirce," in *The Stanford Encyclopedia of Philosophy*, ed. Edward N. Zalta, Winter 2021 (Metaphysics Research Lab, Stanford University, 2021), https://plato.stanford.edu/archives/win2021/entries/peirce/.

10. C. Spearman, "'General Intelligence,' Objectively Determined and Measured," *American Journal of Psychology* 15, no. 2 (April 1904): 284, https://doi.org/10.2307/1412107.

11. The originally identified specific intelligences included linguistic, bodily-kinesthetic, logical-mathematical, musical, spatial, interpersonal, and intrapersonal intelligences: Howard Gardner, *Frames of Mind: The Theory of Multiple Intelligences* (New York: Basic Books, 1983). The dearth of empirical evidence to support multiple intelligences theory has done little to curb its popularity and continued impact on educational and social policy: Lynn Waterhouse, "Multiple Intelligences, the Mozart Effect, and Emotional Intelligence: A Critical Review," *Educational Psychologist* 41, no. 4 (December 1, 2006): 207–225, https://doi.org/10.1207/s15326985ep4104_1.

12. Ian J. Deary, Lars Penke, and Wendy Johnson, "The Neuroscience of Human Intelligence Differences," *Nature Reviews Neuroscience* 11, no. 3 (March 2010): 201–211, https://doi.org/10.1038/nrn2793; Timothy A. Salthouse, "Localizing Age-Related Individual Differences in a Hierarchical Structure," *Intelligence* 32, no. 6 (November–December 2004), https://doi.org/10.1016/j.intell.2004.07.003.

13. Raymond B. Cattell, "Theory of Fluid and Crystallized Intelligence: A Critical Experiment.," *Journal of Educational Psychology* 54, no. 1 (1963): 1–22, https://doi.org/10.1037/h0046743; for the scoop behind the formulation of these ideas, see Richard E. Brown, "Hebb and Cattell: The Genesis of the Theory of Fluid and Crystallized Intelligence," *Frontiers in Human Neuroscience* 10 (December 15, 2016), https://www.frontiersin.org/article/10.3389/fnhum.2016.00606.

14. Raymond B. Cattell, "The Measurement of Adult Intelligence," *Psychological Bulletin* 40, no. 3 (1943): 178, https://doi.org/10.1037/h0059973.

15. Alfred Binet and Théodore Simon, "New Methods for the Diagnosis of the Intellectual Level of Subnormals." *(L'Année Psych.*, 1905, 191–244), in *The Development of Intelligence in Children (The Binet-Simon Scale)* (Baltimore, MD: Williams & Wilkins Co, 1916), 37–90, https://doi.org/10.1037/11069-002.

16. William Stern, *The Psychological Methods of Testing Intelligence*, trans. G. M. Whipple (Baltimore: Warwick & York, 1914).

17. The pattern of a normal distribution of scores is derived from the scores of a norm group, i.e., a large sample of randomly selected individuals who completed the test.

18. Alfred Binet, *Les Idées Modernes Sur Les Enfants* (Paris: Ernest Flammarion, 1909); cited in Stephen Jay Gould, *The Mismeasure of Man* (New York: Norton, 1981).

19. Stuart J. Ritchie and Elliot M. Tucker-Drob, "How Much Does Education Improve Intelligence? A Meta-analysis," *Psychological Science* 29, no. 8 (2018): 1358–1369, https://doi.org/10.1177/0956797618774253. This is an example of a study that shows IQ increases in

the order of one standard deviation with training: Sarah Cassidy et al., "A Relational Frame Skills Training Intervention to Increase General Intelligence and Scholastic Aptitude," *Learning and Individual Differences* 47 (April 2016): 222–235, https://doi.org/10.1016/j.lindif .2016.03.001. Interestingly, IQ scores on average have increased in a sustained manner with each successive generation in many parts of the world over the twentieth century. This is called the Flynn effect: James R. Flynn, *Are We Getting Smarter? Rising IQ in the Twenty-First Century* (Cambridge; New York: Cambridge University Press, 2012). Is there evidence for a commensurate increase in creativity with each passing generation? This question has yet to be empirically examined.

20. Serge Nicolas et al., "Sick? Or Slow? On the Origins of Intelligence as a Psychological Object," *Intelligence* 41, no. 5 (September–October 2013): 699–711, https://doi.org/10.1016/j.intell.2013.08.006.

21. These include the "Synthesis of Three Words in One Sentence" exercise, which is described as "a test in spontaneity, facility of invention and combination"; also, the "Exercise upon Rhymes," which requires imagination and spontaneity; and the assessment of "absurdities" in the "Definitions of Abstract Terms" exercise: Binet and Simon, "New Methods for the Diagnosis of the Intellectual Level of Subnormals."

22. Lewis M. Terman, *The Measurement of Intelligence; An Explanation of and a Complete Guide for the Use of the Stanford Revision and Extension of the Binet-Simon Intelligence Scale* (Boston: Houghton Mifflin Company, 1916), 12.

23. Francis Galton, *Hereditary Genius: An Inquiry into Its Laws and Consequences* (London: Macmillan and Co, 1869), 24, https://doi.org /10.1037/13474-000.

24. Lewis M. Terman and Melita H. Oden, *The Gifted Child Grows Up: Twenty-Five Years' Follow-up of a Superior Group* (Palo Alto, CA: Stanford University Press, 1947), 7.

25. Mitchell Leslie, "The Vexing Legacy of Lewis Terman," *Stanford Magazine*, July 1, 2000, https://stanfordmag.org/contents/the-vexing -legacy-of-lewis-terman.

26. Terman and Oden, *The Gifted Child Grows Up*, 352.

27. Prominent examples include Donald Mackinnon, Ellis Paul Torrance, and Calvin Taylor: Donald W. MacKinnon, "What Makes a Person Creative?," *Theory Into Practice* 5, no. 4 (1966): 152–156; E. Paul Torrance, "Creativity Testing in Education," *Creative Child & Adult Quarterly* 1, no. 3 (1976): 136–148; Calvin W. Taylor and John L. Holland, "Chapter VIII: Development and Application of Tests of Creativity," *Review of Educational Research* 32, no. 1 (February 1962): 91–102, https://doi.org/10.3102/00346543032001091.

28. L. L. Thurstone, "Creative Talent," in *Applications of Psychology* (Oxford: Harper & Brothers, 1952), x, 209.

29. F. Barron, "Originality in Relation to Personality and Intellect," *Journal of Personality* 25, no. 6 (December 1957): 739, https://doi.org /10.1111/j.1467-6494.1957.tb01561.x.

30. J. W. Getzels and P. W. Jackson, *Creativity and Intelligence: Explorations with Gifted Students* (New York: Wiley, 1962).

31. Unlike the Getzels and Jackson study that included children of a variety of ages (sixth graders to high school seniors), Wallach and Kogan's study followed more sound scientific protocols. They examined 150 fifth graders who were relegated to one of four groups based on whether they tested high or low in creativity and intelligence: M. A. Wallach and N. Kogan, *Modes of Thinking in Young Children: A Study of the Creativity-Intelligence Distinction* (New York: Holt, Rinehart & Winston, 1965).

32. Guilford, "Creativity," 445.

33. J. P. Guilford, "Creativity: A Quarter Century of Progress," in *Perspectives in Creativity*, ed. I. A. Taylor and Jacob W. Getzels (Chicago: Aldine, 1975), 38.

34. J. P. Guilford, R. C. Wilson, and P. R. Christensen, "A Factor-Analytic Study of Creative Thinking: Administration of Tests and Analysis of Results. Studies of Aptitudes of High-Level Personnel. II," Reports from the Psychological Laboratory (University of Southern California, July 1952).

35. R. M. Berger, J. P. Guilford, and P. R. Christensen, "A Factor-Analytic Study of Planning Abilities," *Psychological Monographs: General and Applied* 71, no. 6 (1957): 1–31, https://doi.org/10.1037/h0093704.

36. Chief among these is the Torrance Tests of Creative Thinking (TTCT), which was derived from Guilford's work. The TTCT was first published in 1966 and has been re-normed approximately every ten to fifteen years. It remains a highly influential assessment in many parts of the United States within the context of educational practice.

37. J. P. Guilford et al., *Alternate Uses Manual* (Beverly Hills, CA: Sheridan Supply Co., 1960).

38. Early critiques of not just Guilford's work but also that of others like Getzels and Jackson, *Creativity and Intelligence*, include P. E. Vernon, "Creativity and Intelligence," *Educational Research* 6, no. 3 (1964): 163–196, https://doi.org/10.1080/0013188640060301; and A. B. Madans, "What Is Creativity?: II. Some Approaches to the Problem of Creativity," *Transactions of the New York Academy of Sciences* 26, no. 7 (May 1964): 781–787, https://doi.org/10.1111/j.2164-0947.1964.tb01947.x.

39. Robert L. Thorndike, "The Measurement of Creativity," *Teachers College Record* 64, no. 5 (February 1963): 422, https://doi.org/10.1177/016146816306400508.

40. Haier, *The Neuroscience of Intelligence*.

41. For scathing contemporary criticisms of the use of divergent thinking tasks in assessments of creativity, see Arne Dietrich, *How Creativity Happens in the Brain* (Houndmills, Basingstoke, Hampshire; New York: Palgrave Macmillan, 2015); and John Baer, *There's No Such Thing as Creativity: How Plato and 20th Century Psychology Have Misled Us*, Elements in Creativity and Imagination (Cambridge: Cambridge University Press, 2022).

42. Robert J. Sternberg and Linda A. O'Hara, "Creativity and Intelligence," in *Handbook of Creativity*, ed. Robert J. Sternberg (Cambridge: Cambridge University Press, 1999), 611–630.

43. Drevdahl, "Factors of Importance for Creativity," 25.

44. The threshold theory/hypothesis is typically credited in the literature to scholars like Frank Barron and E. Paul Torrance. However, the paper by Drevdahl illustrates that the idea has been around in the individual differences scholarship at least since the early 1950s: Frank Barron, *Creative Person and Creative Process* (Oxford: Holt, Rinehart, & Winston, 1969); E. Paul Torrance, *Guiding Creative Talent* (Englewood Cliffs, NJ: Prentice-Hall, Inc, 1962), https://doi.org/10.1037/13134 -000; Kaoru Yamamoto, "Threshold of Intelligence in Academic Achievement of Highly Creative Students," *Journal of Experimental Education* 32, no. 4 (June 1, 1964): 401–405, https://doi.org/10.1080 /00220973.1964.11010849.

45. Jonathan A. Plucker, Maciej Karwowski, and James C. Kaufman, "Intelligence and Creativity," in *The Cambridge Handbook of Intelligence*, 2nd ed. (New York: Cambridge University Press, 2020), 1087–1105, https://doi.org/10.1017/9781108770422.046.

46. Kyung Hee Kim, "Can Only Intelligent People Be Creative? A Meta-analysis," *Journal of Secondary Gifted Education* 16, no. 2–3 (February 2005): 57–66, https://doi.org/10.4219/jsge-2005-473.

47. Selina Weiss et al., "A Reappraisal of the Threshold Hypothesis of Creativity and Intelligence," *Journal of Intelligence* 8, no. 4 (November 11, 2020): E38, https://doi.org/10.3390/jintelligence8040038.

48. For an accessible and brief overview, see Plucker, Karwowski, and Kaufman, "Intelligence and Creativity."

49. The study actually included an additional measure of originality. This second originality measure was an assessment of the creative ideas that the participants themselves picked as being their best two ideas of all the ones they had generated. An IQ threshold of 104 was found in relation to this top-2 originality measure: Emanuel Jauk et al., "The Relationship between Intelligence and Creativity: New Support for the Threshold Hypothesis by Means of Empirical Breakpoint Detection," *Intelligence* 41, no. 4 (July–August 2013): 212–221, https:// doi.org/10.1016/j.intell.2013.03.003.

50. This review discusses the importance of considering domain specificity: Mark Batey and Adrian Furnham, "Creativity, Intelligence,

and Personality: A Critical Review of the Scattered Literature," *Genetic, Social, and General Psychology Monographs* 132, no. 4 (November 2006): 355–429. It notes, for instance, how MacKinnon found no relationship between creativity and intelligence in a sample of architects, but Barron, *Creativity and Psychological Health*, did in a sample of military officers. See also D. W. MacKinnon, "The Nature and Nurture of Creative Talent," *American Psychologist* 17, no. 7 (1962): 484–495, https://doi.org/10.1037/h0046541.

51. Richard Kazelskis et al., "Two Alternative Definitions of Creativity and Their Relationships with Intelligence," *Journal of Experimental Education* 41, no. 1 (1972): 58–62.

52. For an overview, see Paul J. Silvia, "Intelligence and Creativity Are Pretty Similar after All," *Educational Psychology Review* 27, no. 4 (2015): 599–606, https://doi.org/10.1007/s10648-015-9299-1.

53. Allison C. Sligh, Frances A. Conners, and Beverly Roskos-Ewoldsen, "Relation of Creativity to Fluid and Crystallized Intelligence," *Journal of Creative Behavior* 39, no. 2 (2005): 123–136, https://doi.org/10.1002/j.2162-6057.2005.tb01254.x.

54. Silvia, "Intelligence and Creativity Are Pretty Similar after All," 603.

55. David Lubinski, Camilla P. Benbow, and Harrison J. Kell, "Life Paths and Accomplishments of Mathematically Precocious Males and Females Four Decades Later," *Psychological Science* 25, no. 12 (2014): 2217–2232, https://doi.org/10.1177/0956797614551371.

56. Barron, "Originality in Relation to Personality and Intellect," 735.

57. Robert J. Sternberg and Todd I. Lubart, "An Investment Theory of Creativity and Its Development," *Human Development* 34, no. 1 (1991): 1–31, https://doi.org/10.1159/000277029; Robert J. Sternberg, "The Theory of Successful Intelligence," *Review of General Psychology* 3, no. 4 (1999): 292–316, https://doi.org/10.1037/1089-2680.3.4.292; Robert J. Sternberg, "A Triangular Theory of Creativity," *Psychology of Aesthetics, Creativity, and the Arts* 12, no. 1 (2018): 50–67, https://doi.org/10.1037/aca0000095; John Baer and James C. Kaufman, "Bridging

Generality and Specificity: The Amusement Park Theoretical (APT) Model of Creativity," *Roeper Review* 27, no. 3 (2005): 158–163, https:// doi.org/10.1080/02783190509554310.

58. W. E. B. Du Bois, "Criteria of Negro Art," in *The New Negro: Readings on Race, Representation, and African American Culture, 1892–1938*, ed. Henry Louis Gates Jr. and Gene Andrew Jarrett (Princeton, NJ: Princeton University Press, [1926] 2007), 257–260; 258.

59. As noted in Lev Semenovich Vygotsky, *The Collected Works of L. S. Vygotsky, Vol. 5: Child Psychology* (New York: Plenum Press, 1998), 4.

60. Albeit rarely in juxtaposition with intelligence as an added variable for consideration: McCrae, "Creativity, Divergent Thinking, and Openness to Experience"; Beth A. Hennessey and Teresa M. Amabile, "The Conditions of Creativity," in *The Nature of Creativity: Contemporary Psychological Perspectives*, ed. Robert J. Sternberg (Cambridge: Cambridge University Press, 1988), 11–43. For examples of studies that have considered motivation and interest (as assessed by openness to experience) in relation to the creativity intelligence link, see Alexandra M. Harris, Rachel L. Williamson, and Nathan T. Carter, "A Conditional Threshold Hypothesis for Creative Achievement: On the Interaction between Intelligence and Openness," *Psychology of Aesthetics, Creativity, and the Arts* 13, no. 3 (2019): 322–337, https:// doi.org/10.1037/aca0000182; and Chee-Seng Tan et al., "Openness to Experience Enhances Creativity: The Mediating Role of Intrinsic Motivation and the Creative Process Engagement," *Journal of Creative Behavior* 53, no. 1 (March 2019): 109–119, https://doi.org/10.1002 /jocb.170.

61. An exception is Alice W. Flaherty, "Homeostasis and the Control of Creative Drive," in *The Cambridge Handbook of the Neuroscience of Creativity* (New York: Cambridge University Press, 2018), 19–49, https://doi.org/10.1017/9781316556238.003.

62. Théodule Ribot, *Essay on the Creative Imagination* (Chicago: Open Court Publishing Company, 1906), 8.

63. Andrew Robinson, *The Story of Measurement* (London; New York: Thames & Hudson, 2007), 7.

64. Thomas S. Kuhn, *The Structure of Scientific Revolutions*, 2nd ed. (Chicago: University of Chicago Press, 1970).

65. Theodore M. Porter, *Trust in Numbers: The Pursuit of Objectivity in Science and Public Life* (Princeton, NJ: Princeton University Press, 1995), 8.

Chapter 6

1. *Dr. Oliver Sacks, Academy Class of 2000, Part 10*, 2017, https://youtu .be/e4kCpqvUG6A; for more in relation to the spectacular impact of L-DOPA, see Oliver Sacks, *Awakenings* (Garden City, NY: Doubleday, 1974). L-DOPA or levodopa is a precursor to catecholamines (the neurotransmitter family of dopamine, norepinephrine, and epinephrine). As L-DOPA can cross the blood-brain barrier and is converted into dopamine in the central nervous system, it is used in the clinical treatment of Parkinson's disease and other motor disorders.

2. J. F. Keeler, D. O. Pretsell, and T. W. Robbins, "Functional Implications of Dopamine D1 vs. D2 Receptors: A 'Prepare and Select' Model of the Striatal Direct vs. Indirect Pathways," in "The Ventral Tegmentum and Dopamine: A New Wave of Diversity," ed. Michel Barrot, special issue, *Neuroscience* 282 (December 12, 2014): 169, https://doi .org/10.1016/j.neuroscience.2014.07.021.

3. Richard J. Beninger, *Life's Rewards: Linking Dopamine, Incentive Learning, Schizophrenia, and the Mind* (New York: Oxford University Press, 2018), 50.

4. Terrence Sejnowski, "Dopamine Made You Do It," in *Think Tank: Forty Neuroscientists Explore the Biological Roots of Human Experience*, ed. David J. Linden (New Haven: Yale University Press, 2018), 262, https://doi.org/10.12987/9780300235470.

5. Vaughan Bell, "The Unsexy Truth about Dopamine," *Observer*, February 3, 2013, sec. Science, https://www.theguardian.com/science /2013/feb/03/dopamine-the-unsexy-truth.

6. Here are some noteworthy examples of articles that make a case for dopamine as the cause of the world's problems, and explain how to

avoid dopamine to improve wellbeing and also how to boost dopamine to feel better: Nellie Bowles, "How to Feel Nothing Now, in Order to Feel More Later," *New York Times*, November 7, 2019, sec. Style, https://www.nytimes.com/2019/11/07/style/dopamine-fasting .html; Steven Kotler, "Addicted to Bang: The Neuroscience of the Gun," *Forbes*, December 18, 2012, https://www.forbes.com/sites/steven kotler/2012/12/18/addicted-to-bang-the-neuroscience-of-the-gun/; Erica Julson, "10 Best Ways to Increase Dopamine Levels Naturally," Healthline, February 28, 2022, https://www.healthline.com/nutrition /how-to-increase-dopamine.

7. Javier Cuevas, "Neurotransmitters and Their Life Cycle☆," in *Reference Module in Biomedical Sciences* (Elsevier, 2019), https://doi.org /10.1016/B978-0-12-801238-3.11318-2.

8. Estimations for other amine neurotransmitters like serotonin and noradrenaline are also 1 percent. This is in contrast to GABA, the major inhibitory neurotransmitter, where the estimation of synapses in the brain is 30–40 percent: Solomon H. Snyder, "The Brain Harbors Many Neurotransmitters," in *Think Tank: Forty Neuroscientists Explore the Biological Roots of Human Experience* (Yale University Press, 2018), 88–94, https://jhu.pure.elsevier.com/en/publications/the-brain-harbors -many-neurotransmitters. It is clear that the future will bring the discovery of several more neurotransmitters, but what remains a mystery is estimating how many there will be in all.

9. The domain-overlap model has been proposed to explain the distinct pattern associated with dopamine neurotransmission: Changliang Liu, Pragya Goel, and Pascal S. Kaeser, "Spatial and Temporal Scales of Dopamine Transmission," *Nature Reviews Neuroscience* 22, no. 6 (2021): 345–358, https://doi.org/10.1038/s41583-021-00455-7.

10. For a great overview on the topic, watch this lecture by Richard Palmiter, professor of biochemistry and Howard Hughes Medical Institute (HHMI) investigator: *Dopamine: Motivation, Salience and Learning—A Genetic Perspective*, 2010, https://youtu.be/bVN6426oVgw.

11. Roy A. Wise, "Roles for Nigrostriatal—Not Just Mesocorticolimbic —Dopamine in Reward and Addiction," *Trends in Neurosciences* 32,

no. 10 (October 2009): 517–524, https://doi.org/10.1016/j.tins.2009 .06.004; Robert A. McCutcheon, Anissa Abi-Dargham, and Oliver D. Howes, "Schizophrenia, Dopamine and the Striatum: From Biology to Symptoms," *Trends in Neurosciences* 42, no. 3 (March 2019): 205–220, https://doi.org/10.1016/j.tins.2018.12.004.

12. James Olds and Peter Milner, "Positive Reinforcement Produced by Electrical Stimulation of Septal Area and Other Regions of Rat Brain," *Journal of Comparative and Physiological Psychology* 47, no. 6 (1954): 419–427, https://doi.org/10.1037/h0058775. The rewarding effect of electrical self-stimulation in the septal area and other regions was corroborated in studies on humans as well: M. P. Bishop, S. Thomas Elder, and Robert G. Heath, "Intracranial Self-Stimulation in Man," *Science* 140, no. 3565 (1963): 394–396.

13. James Olds, "Pleasure Centers in the Brain," *Scientific American* 195, no. 4 (1956): 116.

14. While the noradrenaline/norepinephrine system was also found to be involved, dopamine was regarded as more relevant to reward processing. "Both the norepinephrine and the dopamine systems overlap much of the area that gives rise to self-stimulation behavior in rats. The dopamine fibers are found only in areas that mediate brain reward, whereas the norepinephrine fibers extend into other regions. This and other evidence points [sic] to a more critical role for dopamine in brain reward": Aryeh Routtenberg, "The Reward System of the Brain," *Scientific American* 239, no. 5 (November 1978): 159, https://doi.org/10.1038/scientificamerican1178-154.

15. Eric J. Nestler, "Is There a Common Molecular Pathway for Addiction?," *Nature Neuroscience* 8, no. 11 (2005): 1445–1449, https://doi .org/10.1038/nn1578; Roy A. Wise, "Drug-Activation of Brain Reward Pathways," *Drug and Alcohol Dependence* 51, no. 1–2 (June 1998): 13–22, https://doi.org/10.1016/S0376-8716(98)00063-5.

16. Kent C. Berridge and Morten L. Kringelbach, "Affective Neuroscience of Pleasure: Reward in Humans and Animals," *Psychopharmacology* 199, no. 3 (August 2008): 457–480, https://doi.org/10.1007/s00 213-008-1099-6.

17. Matthijs Baas, Carsten K. W. De Dreu, and Bernard A. Nijstad, "A Meta-analysis of 25 Years of Mood-Creativity Research: Hedonic Tone, Activation, or Regulatory Focus?," *Psychological Bulletin* 134, no. 6 (November 2008): 779–806, https://doi.org/10.1037/a0012815; Mark A. Davis, "Understanding the Relationship between Mood and Creativity: A Meta-analysis," *Organizational Behavior and Human Decision Processes* 108, no. 1 (January 2009): 25–38, https://doi.org/10.1016/j .obhdp.2008.04.001.

18. Gesine Dreisbach and Thomas Goschke, "How Positive Affect Modulates Cognitive Control: Reduced Perseveration at the Cost of Increased Distractibility.," *Journal of Experimental Psychology: Learning, Memory, and Cognition* 30, no. 2 (March 2004): 343–353, https://doi .org/10.1037/0278-7393.30.2.343; F. Gregory Ashby, Vivian V. Valentin, and And U. Turken, "Chapter 11—The Effects of Positive Affect and Arousal and Working Memory and Executive Attention: Neurobiology and Computational Models," in *Emotional Cognition: From Brain to Behaviour*, ed. Simon C. Moore and Mike Oaksford, Advances in Consciousness Research, vol. 44 (Amsterdam: John Benjamins Publishing Company, 2002), 245–287, https://doi.org/10.1075/aicr .44.11ash.

19. Improvements were found on tasks of creative problem-solving as well as convergent creative thinking: Alice M. Isen, Kimberly A. Daubman, and Gary P. Nowicki, "Positive Affect Facilitates Creative Problem Solving," *Journal of Personality and Social Psychology* 52, no. 6 (1987): 1122–1131, https://doi.org/10.1037/0022-3514.52.6.1122.

20. Tanis Bryan and James Bryan, "Positive Mood and Math Performance," *Journal of Learning Disabilities* 24, no. 8 (1991): 490–494, https://doi.org/10.1177/002221949102400808.

21. Alice M. Isen, "The Influence of Positive and Negative Affect on Cognitive Organization: Some Implications for Development," in *Psychological and Biological Approaches to Emotion* (Hillsdale, NJ: Lawrence Erlbaum Associates, Inc, 1990), 75–94.

22. Beth A Hennessey and Teresa M Amabile, "Creativity," *Annual Review of Psychology* 61 (2010): 581, https://doi.org/10.1146/annurev

.psych.093008.100416. This view, however, has not gone unchallenged: R. Eisenberger and J. Cameron, "Detrimental Effects of Reward. Reality or Myth?," *American Psychologist* 51, no. 11 (1996): 1153–1166, https://doi.org/10.1037//0003-066x.51.11.1153. The reality in fact is more nuanced. A meta-analysis of the rewards-creativity association in children and adults has revealed that performance-contingent rewards have a modest negative effect on creative performance whereas creativity-contingent rewards tend to have a conducive influence: Kris Byron and Shalini Khazanchi, "Rewards and Creative Performance: A Meta-analytic Test of Theoretically Derived Hypotheses," *Psychological Bulletin* 138, no. 4 (July 2012): 809–830, https://doi.org/10.1037/a0027652.

23. For instance, a recent study demonstrated that dopamine associates effect with cause rather than the opposite (which is what is typically assumed in reward-prediction models): Huijeong Jeong et al., "Mesolimbic Dopamine Release Conveys Causal Associations," *Science* 378, no. 6626 (December 23, 2022): eabq6740, https://doi.org /10.1126/science.abq6740.

24. Unpredictability in fact seems to be a key factor that elicits dopamine activity: Wolfram Schultz, Peter Dayan, and P. Read Montague, "A Neural Substrate of Prediction and Reward," *Science* 275, no. 5306 (March 14, 1997): 1593–1599, https://doi.org/10.1126/science.275 .5306.1593; J. Mirenowicz and W. Schultz, "Importance of Unpredictability for Reward Responses in Primate Dopamine Neurons," *Journal of Neurophysiology* 72, no. 2 (August 1994): 1024–1027, https://doi.org /10.1152/jn.1994.72.2.1024.

25. T. Ljungberg, P. Apicella, and W. Schultz, "Responses of Monkey Dopamine Neurons during Learning of Behavioral Reactions," *Journal of Neurophysiology* 67, no. 1 (January 1992): 145–163, https://doi.org /10.1152/jn.1992.67.1.145.

26. Masayuki Matsumoto and Okihide Hikosaka, "Two Types of Dopamine Neuron Distinctly Convey Positive and Negative Motivational Signals," *Nature* 459 (2009): 837–841, https://doi.org/10.1038 /nature08028.

27. Jon C. Horvitz, Tripp Stewart, and Barry L. Jacobs, "Burst Activity of Ventral Tegmental Dopamine Neurons Is Elicited by Sensory Stimuli in the Awake Cat," *Brain Research* 759, no. 2 (June 6, 1997): 251–258, https://doi.org/10.1016/S0006-8993(97)00265-5.

28. Wolfram Schultz, "Behavioral Dopamine Signals," in "Fifty Years of Dopamine Research," ed. Anders Björklund and Stephen B. Dunnett, special issue, *Trends in Neurosciences* 30, no. 5 (May 2007): 203, https://doi.org/10.1016/j.tins.2007.03.007.

29. Mark D'Esposito and Bradley R. Postle, "The Cognitive Neuroscience of Working Memory," *Annual Review of Psychology* 66, no. 1 (2015): 115–142, https://doi.org/10.1146/annurev-psych-010814-015031.

30. Torben Ott and Andreas Nieder, "Dopamine and Cognitive Control in Prefrontal Cortex," *Trends in Cognitive Sciences* 23, no. 3 (March 2019): 213–234, https://doi.org/10.1016/j.tics.2018.12.006.

31. Andrew Westbrook, Michael J. Frank, and Roshan Cools, "A Mosaic of Cost–Benefit Control over Cortico-Striatal Circuitry," *Trends in Cognitive Sciences* 25, no. 8 (August 2021): 710–721, https://doi.org/10.1016/j.tics.2021.04.007.

32. Nathalie Boot et al., "Creative Cognition and Dopaminergic Modulation of Fronto-Striatal Networks: Integrative Review and Research Agenda," *Neuroscience & Biobehavioral Reviews* 78 (July 2017): 13–23, https://doi.org/10.1016/j.neubiorev.2017.04.007. For a recent expansion of this view, see Ceyda Sayalı and Frederick S. Barrett, "The Costs and Benefits of Psychedelics on Cognition and Mood," *Neuron* 11, no. 5 (March 2023), 614–630, https://doi.org/10.1016/j.neuron.2022.12.031.

33. Craig N. Karson, "Spontaneous Eye-Blink Rates and Dopaminergic Systems," *Brain: A Journal of Neurology* 106, no. 3 (September 1983): 643–653, https://doi.org/10.1093/brain/106.3.643.

34. Eye-blink rates were negatively correlated with convergent thinking and showed an inverted-U relation with divergent thinking: Soghra Akbari Chermahini and Bernhard Hommel, "The (b)Link between Creativity and Dopamine: Spontaneous Eye Blink Rates

Predict and Dissociate Divergent and Convergent Thinking," *Cognition* 115, no. 3 (June 2010): 458–465, https://doi.org/10.1016/j.cognition.2010.03.007; additional support from Sergio Agnoli et al., "Dopamine Supports Idea Originality: The Role of Spontaneous Eye Blink Rate on Divergent Thinking," *Psychological Research* 87 (2023), 17–23, https://doi.org/10.1007/s00426-022-01658-y.

35. Stan B. Floresco, "Prefrontal Dopamine and Behavioral Flexibility: Shifting from an 'Inverted-U' toward a Family of Functions," *Frontiers in Neuroscience* 7 (April 19, 2013), https://www.frontiersin.org/article/10.3389/fnins.2013.00062; Kimberlee D'Ardenne et al., "Role of Prefrontal Cortex and the Midbrain Dopamine System in Working Memory Updating," *Proceedings of the National Academy of Sciences of the United States of America* 109, no. 49 (December 4, 2012): 19900–19909, https://doi.org/10.1073/pnas.1116727109; T. W Robbins, "Shifting and Stopping: Fronto-Striatal Substrates, Neurochemical Modulation and Clinical Implications," *Philosophical Transactions of the Royal Society B* 362 (2007): 917–932, https://doi.org/10.1098/rstb.2007.2097.

36. Daniel Durstewitz and Jeremy K. Seamans, "The Dual-State Theory of Prefrontal Cortex Dopamine Function with Relevance to Catechol-O-Methyltransferase Genotypes and Schizophrenia," *Biological Psychiatry*, Neurodevelopment and the Transition from Schizophrenia Prodrome to Schizophrenia, 64, no. 9 (November 2008): 739, https://doi.org/10.1016/j.biopsych.2008.05.015.

37. Jeremy K. Seamans and Trevor W. Robbins, "Dopamine Modulation of the Prefrontal Cortex and Cognitive Function," in *The Dopamine Receptors*, ed. Kim A. Neve (Totowa, NJ: Humana Press, 2010), 373–398, https://doi.org/10.1007/978-1-60327-333-6_14.

38. A modest positive effect was found for a convergent creative-thinking task that involved generating remote associates to reach a single correct response: Lorenza S. Colzato, Annelies M. de Haan, and Bernhard Hommel, "Food for Creativity: Tyrosine Promotes Deep Thinking," *Psychological Research*, September 26, 2014, https://doi.org/10.1007/s00426-014-0610-4. See note 34 for an example of the opposite pattern of finding from the same research group.

39. Shawn F. Smyth and David Q. Beversdorf, "Lack of Dopaminergic Modulation of Cognitive Flexibility," *Cognitive and Behavioral Neurology: Official Journal of the Society for Behavioral and Cognitive Neurology* 20, no. 4 (December 2007): 225–229, https://doi.org/10.1097/WNN.0b013e31815e6244.

40. Low latent inhibition or "the '   to screen out previously irrelevant stimuli" is positively associated with creative achievement: Shelley H. Carson, Jordan B. Peterson, and Daniel M. Higgins, "Decreased Latent Inhibition Is Associated with Increased Creative Achievement in High-Functioning Individuals," *Journal of Personality and Social Psychology* 85, no. 3 (September 2003): 499, https://doi.org/10.1037/0022-3514.85.3.499. In a similar vein, creative achievement is associated with "leaky attention" rather than "flexible attention": Darya L. Zabelina, Arielle Saporta, and Mark Beeman, "Flexible or Leaky Attention in Creative People? Distinct Patterns of Attention for Different Types of Creative Thinking," *Memory & Cognition* 44, no. 3 (April 2016): 488–498, https://doi.org/10.3758/s13421-015-0569-4.

41. Martine Hoogman et al., "Creativity and ADHD: A Review of Behavioral Studies, the Effect of Psychostimulants and Neural Underpinnings," *Neuroscience & Biobehavioral Reviews* 119 (December 2020): 66–85, https://doi.org/10.1016/j.neubiorev.2020.09.029.

42. A parallel case is made for noradrenaline/norepinephrine: Kenneth M. Heilman, Stephen E. Nadeau, and David O. Beversdorf, "Creative Innovation: Possible Brain Mechanisms," *Neurocase* 9, no. 5 (2003): 369–379, https://doi.org/10.1076/neur.9.5.369.16553; Molly McBride et al., "Effects of Stimulant Medication on Divergent and Convergent Thinking Tasks Related to Creativity in Adults with Attention-Deficit Hyperactivity Disorder," *Psychopharmacology* 238, no. 12 (2021): 3533–3541, https://doi.org/10.1007/s00213-021-05970-0.

43. Roshan Cools et al., "Chapter 7—Dopamine and the Motivation of Cognitive Control," in *The Frontal Lobes, Handbook of Clinical Neurology*, ed. Mark D'Esposito and Jordan H. Grafman, Handbook of Clinical Neurology, vol. 163 (Elsevier, 2019), 123–143, https://doi.org/10.1016/B978-0-12-804281-6.00007-0.

44. Andrew Westbrook and Todd S. Braver, "Dopamine Does Double Duty in Motivating Cognitive Effort," *Neuron* 89, no. 4 (February 17, 2016): 695, https://doi.org/10.1016/j.neuron.2015.12.029.

45. For more information on this view of the dopamine-creativity link: Flaherty, "Brain Illness and Creativity"; Flaherty, "Frontotemporal and Dopaminergic Control of Idea Generation and Creative Drive"; Flaherty, "Homeostasis and the Control of Creative Drive"; Flaherty, *The Midnight Disease*.

46. Rex E. Jung et al., "Quantity Yields Quality When It Comes to Creativity: A Brain and Behavioral Test of the Equal-Odds Rule," *Frontiers in Psychology* 6 (2015), https://doi.org/10.3389/fpsyg.2015.00864.

47. The "ideas" generated in this study are best conceived as visuomotor options: Yuen-Siang Ang et al., "Dopamine Modulates Option Generation for Behavior," *Current Biology* 28, no. 10 (May 21, 2018): 1561–1569.e3, https://doi.org/10.1016/j.cub.2018.03.069.

48. Pedro J. Garcia-Ruiz, Juan Carlos Martinez Castrillo, and Lydia Vela Desojo, "Creativity Related to Dopaminergic Treatment: A Multicenter Study," *Parkinsonism & Related Disorders* 63 (June 2019): 169–173, https://doi.org/10.1016/j.parkreldis.2019.02.010; Eugénie Lhommée et al., "Dopamine and the Biology of Creativity: Lessons from Parkinson's Disease," *Frontiers in Neurology* 5 (2014): 55, https://doi.org/10.3389/fneur.2014.00055; Rivka Inzelberg, "The Awakening of Artistic Creativity and Parkinson's Disease," *Behavioral Neuroscience* 127, no. 2 (2013): 256–261, https://doi.org/10.1037/a0031052.

49. Margherita Canesi et al., "Creative Thinking, Professional Artists, and Parkinson's Disease," *Journal of Parkinson's Disease* 6, no. 1 (2016): 239–246, https://doi.org/10.3233/JPD-150681.

50. Lauring et al., "Why Would Parkinson's Disease Lead to Sudden Changes in Creativity, Motivation, or Style with Visual Art?"

51. Indeed, this paper makes a case for channeling the release of energies that come from ingesting dopamine agonists to productive activities—such as artistic endeavors—by providing a positive environment for the same: Pedro J. Garcia-Ruiz, "Impulse Control Disorders

and Dopamine-Related Creativity: Pathogenesis and Mechanism, Short Review, and Hypothesis," *Frontiers in Neurology* 9 (2018): 1041, https://doi.org/10.3389/fneur.2018.01041.

52. Characterized as dopaminergic hyperinnervation: Tiago V. Maia and Vasco A. Conceição, "Dopaminergic Disturbances in Tourette Syndrome: An Integrative Account," *Biological Psychiatry* 84, no. 5 (September 2018): 332–344, https://doi.org/10.1016/j.biopsych.2018.02.1172.

53. Sacks, "Tourette's Syndrome and Creativity."

54. Mei-Hue Wei, "Social Adjustment, Academic Performance, and Creativity of Taiwanese Children with Tourette's Syndrome," *Psychological Reports* 108, no. 3 (2011): 791–798, https://doi.org/10.2466/04.07.10.PR0.108.3.791-798; Carlotta Zanaboni Dina et al., "Creativity Assessment in Subjects with Tourette Syndrome vs. Patients with Parkinson's Disease: A Preliminary Study," *Brain Sciences* 7, no. 7 (July 2017), https://doi.org/10.3390/brainsci7070080.

55. Flaherty, "Homeostasis and the Control of Creative Drive," 24.

56. It is worth noting that at baseline apathetic individuals showed higher uniqueness and low fluency during motoric option generation whereas motivated individuals showed the opposite profile of high fluency and low uniqueness: Ang et al., "Dopamine Modulates Option Generation for Behavior."

57. Hila Z. Gvirts et al., "Novelty-Seeking Trait Predicts the Effect of Methylphenidate on Creativity," *Journal of Psychopharmacology* (Oxford) 31, no. 5 (2017): 599–605, https://doi.org/10.1177/0269881116667703.

58. Martha J. Farah et al., "When We Enhance Cognition with Adderall, Do We Sacrifice Creativity? A Preliminary Study," *Psychopharmacology* 202, no. 1–3 (2009): 541–547, https://doi.org/10.1007/s00213-008-1369-3.

59. Alice Flaherty, *Danger and Creativity*, TEDxAmherst, 2019, https://www.ted.com/talks/alice_flaherty_danger_and_creativity.

60. This parallels influential models of personality: Marvin Zucker-man, *Psychobiology of Personality* (New York: Cambridge University Press, 1991); Marvin Zuckerman and D. Michael Kuhlman, "Personality and Risk-Taking: Common Bisocial Factors," *Journal of Personality* 68, no. 6 (2000): 999–1029, https://doi.org/10.1111/1467-6494.00124.

61. Radwa Khalil and Ahmed A. Moustafa, "A Neurocomputational Model of Creative Processes," *Neuroscience & Biobehavioral Reviews* 137 (June 2022): 104656, https://doi.org/10.1016/j.neubiorev.2022.104656. In contrast, Simeng Gu et al. attribute the role of dopamine to value appraisals in "The Neural Mechanism Underlying Cognitive and Emotional Processes in Creativity," *Frontiers in Psychology* 9 (2018): 1924, https://doi.org/10.3389/fpsyg.2018.01924.

62. Openness was positively associated with the degree of functional connectivity between dopaminergic input regions (substantia nigra and ventral tegmental area) and the dorsolateral prefrontal cortex (DLPFC), an area known for its role in cognitive control in facilitating goal-directed and adaptive thought and action: L. Passamonti et al., "Increased Functional Connectivity within Mesocortical Networks in Open People," *NeuroImage* 104, no. 1 (January 2015): 301–309, https://doi.org/10.1016/j.neuroimage.2014.09.017.

63. Other facets of novelty seeking, such as disorderliness, extravagance, and impulsivity, were not associated with high creative achievement: Rosa Aurora Chavez-Eakle, Ma. del Carmen Lara, and Carlos Cruz-Fuentes, "Personality: A Possible Bridge between Creativity and Psychopathology?," *Creativity Research Journal* 18, no. 1 (2006): 27–38, https://doi.org/10.1207/s15326934crj1801_4.

64. Celeste Kidd and Benjamin Y. Hayden, "The Psychology and Neuroscience of Curiosity," *Neuron* 88, no. 3 (November 4, 2015): 449–460, https://doi.org/10.1016/j.neuron.2015.09.010.

65. Matthias J. Gruber, Bernard D. Gelman, and Charan Ranganath, "States of Curiosity Modulate Hippocampus-Dependent Learning via the Dopaminergic Circuit," *Neuron* 84, no. 2 (October 22, 2014): 486–496, https://doi.org/10.1016/j.neuron.2014.08.060.

66. Jaak Panksepp and Joseph Moskal, "Dopamine and SEEKING: Subcortical 'Reward' Systems and Appetitive Urges," in *Handbook of Approach and Avoidance Motivation* (New York: Psychology Press, 2008), 67–87.

67. Kenneth L. Davis, Jaak Panksepp, and Larry Normansell, "The Affective Neuroscience Personality Scales: Normative Data and Implications," *Neuropsychoanalysis* 5, no. 1 (2003): 59, https://doi.org/10.1080/15294145.2003.10773410.

68. Richard A. Depue and Paul F. Collins, "Neurobiology of the Structure of Personality: Dopamine, Facilitation of Incentive Motivation, and Extraversion," *Behavioral and Brain Sciences* 22, no. 3 (June 1999): 491–517; discussion 518–569, https://doi.org/10.1017/s0140525x99002046.

69. Colin G. DeYoung et al., "Sources of Cognitive Exploration: Genetic Variation in the Prefrontal Dopamine System Predicts Openness/Intellect," *Journal of Research in Personality* 45, no. 4 (August 2011): 364–371, https://doi.org/10.1016/j.jrp.2011.04.002.

70. Colin G. Deyoung and Jeremy R. Gray, "Personality Neuroscience: Explaining Individual Differences in Affect, Behaviour and Cognition," in *The Cambridge Handbook of Personality Psychology* (New York: Cambridge University Press, 2009), 323–346, https://doi.org/10.1017/CBO9780511596544.023.

71. This is distinguished from the metatrait of stability or self-control (from the shared variance of the remaining big five traits: emotional stability, conscientious, agreeableness), which reflects "the need to maintain a stable organization of behavioral and psychological function": Jacob B. Hirsh, Colin G. Deyoung, and Jordan B. Peterson, "Metatraits of the Big Five Differentially Predict Engagement and Restraint of Behavior," *Journal of Personality* 77, no. 4 (August 2009): 1086, https://doi.org/10.1111/j.1467-6494.2009.00575.x.

72. Colin G. DeYoung, "The Neuromodulator of Exploration: A Unifying Theory of the Role of Dopamine in Personality," *Frontiers in Human Neuroscience* 7 (2013): 8, https://doi.org/10.3389/fnhum.2013.00762.

73. DeYoung, "The Neuromodulator of Exploration," 8.

74. Wiebke Käckenmester, Antonia Bott, and Jan Wacker, "Openness to Experience Predicts Dopamine Effects on Divergent Thinking," *Personality Neuroscience* 2 (2019): e3, https://doi.org/10.1017/pen.2019.3.

75. Interestingly, dysfunctions within the dopaminergic systems have been proposed to explain erroneous attribution of salience to irrelevant information in schizophrenia: Shitij Kapur, "Psychosis as a State of Aberrant Salience: A Framework Linking Biology, Phenomenology, and Pharmacology in Schizophrenia," *American Journal of Psychiatry* 160, no. 1 (January 2003): 13–23, https://doi.org/10.1176/appi.ajp.160.1.13.

76. J. C. Horvitz, "Mesolimbocortical and Nigrostriatal Dopamine Responses to Salient Non-reward Events," *Neuroscience* 96, no. 4 (2000): 654, https://doi.org/10.1016/s0306-4522(00)00019-1. There are marked similarities in this position and those advocated in relation to cognitive control (see note 30).

77. Martin Tik et al., "Ultra-High-Field FMRI Insights on Insight: Neural Correlates of the Aha!-Moment," *Human Brain Mapping* 39, no. 8 (August 2018): 3241–3252, https://doi.org/10.1002/hbm.24073. Regions of the salience network in the brain (anterior insula) have also been showed to be engaged during divergent thinking in highly creative individuals when compared to a low creative group: Abraham et al., "Creative Conceptual Expansion."

78. Martin Reuter et al., "Identification of First Candidate Genes for Creativity: A Pilot Study," *Brain Research* 1069, no. 1 (January 2006): 190–197, https://doi.org/10.1016/j.brainres.2005.11.046.

79. Anne Chong et al., "Blending Oxytocin and Dopamine with Everyday Creativity," *Scientific Reports* 11, no. 1 (August 10, 2021): 16185, https://doi.org/10.1038/s41598-021-95724-x.

80. Darya L. Zabelina et al., "Dopamine and the Creative Mind: Individual Differences in Creativity Are Predicted by Interactions between Dopamine Genes DAT and COMT," *PloS ONE* 11, no. 1 (2016): e0146768, https://doi.org/10.1371/journal.pone.0146768.

81. A. N. Kluger, Z. Siegfried, and R. P. Ebstein, "A Meta-analysis of the Association between DRD4 Polymorphism and Novelty Seeking," *Molecular Psychiatry* 7 (2002): 712–717, https://doi.org/10.1038 /sj.mp.4001082.

82. Naama Mayseless et al., "The Association between Creativity and 7R Polymorphism in the Dopamine Receptor D4 Gene (DRD4)," *Frontiers in Human Neuroscience* 7 (2013), https://doi.org/10.3389 /fnhum.2013.00502.

83. So, for instance, by using the neuroimaging technique of diffusion tensor imaging, it is possible to assess brain tissue density by measuring mean diffusivity. This is a metric of the degree to which water molecules are able to move freely within the tissue. Lower mean diffusivity points to higher tissue density: Hikaru Takeuchi and Ryuta Kawashima, "Mean Diffusivity in the Dopaminergic System and Neural Differences Related to Dopaminergic System," *Current Neuropharmacology* 16, no. 4 (2018): 460–474, https://doi.org/10.2174 /1570159X15666171109124839.

84. Hikaru Takeuchi et al., "Mean Diffusivity of Globus Pallidus Associated with Verbal Creativity Measured by Divergent Thinking and Creativity-Related Temperaments in Young Healthy Adults," *Human Brain Mapping* 36, no. 5 (May 2015): 1808–1827, https://doi.org/10.1002 /hbm.22739; Hikaru Takeuchi et al., "Regional Gray Matter Volume of Dopaminergic System Associate[d] with Creativity: Evidence from Voxel-Based Morphometry," *NeuroImage* 51, no. 2 (June 2010): 578–585, https://doi.org/10.1016/j.neuroimage.2010.02.078; Hikaru Takeuchi et al., "White Matter Structures Associated with Creativity: Evidence from Diffusion Tensor Imaging," *NeuroImage* 51, no. 1 (May 15, 2010): 11–18, https://doi.org/10.1016/j.neuroimage.2010.02.035.

85. Hikaru Takeuchi et al., "Creative Females Have Larger White Matter Structures: Evidence from a Large Sample Study," *Human Brain Mapping* 38 (2017), 414–430, https://doi.org/10.1002/hbm.23369.

86. de Manzano et al., "Thinking Outside a Less Intact Box."

87. Rex E. Jung et al., "White Matter Integrity, Creativity, and Psychopathology: Disentangling Constructs with Diffusion Tensor Imaging,"

PloS ONE 5, no. 3 (2010): e9818, https://doi.org/10.1371/journal.pone
.0009818; Wertz et al., "White Matter Correlates of Creative Cognition in a Normal Cohort."

88. Ethan S. Bromberg-Martin, Masayuki Matsumoto, and Okihide Hikosaka, "Dopamine in Motivational Control: Rewarding, Aversive, and Alerting," *Neuron* 68, no. 5 (December 2010): 815–834, https://doi.org/10.1016/j.neuron.2010.11.022.

89. Marc D. Lewis, "Desire, Dopamine, and Conceptual Development," in *Child Development at the Intersection of Emotion and Cognition*, ed. Susan D. Calkins and Martha Ann Bell, Human Brain Development (Washington, DC: American Psychological Association, 2010), 175–199, https://doi.org/10.1037/12059-010.

Chapter 7

1. Joseph Carroll, "Imagination, the Brain's Default Mode Network, and Imaginative Verbal Artifacts," in *Evolutionary Perspectives on Imaginative Culture*, ed. Joseph Carroll, Mathias Clasen, and Emelie Jonsson (Cham: Springer International Publishing, 2020), 31–52, https://doi.org/10.1007/978-3-030-46190-4_2. The phrase "imagination network" and its link to the default mode network originates from Scott Barry Kaufman and Carolyn Gregoire, *Wired to Create: Unraveling the Mysteries of the Creative Mind* (New York: Perigee Books, 2015), xxvii.

2. Sally L. Satel and Scott O. Lilienfeld, *Brainwashed: The Seductive Appeal of Mindless Neuroscience* (New York: Basic Books, 2013), xiv. The term "premature extrapolation" comes from Steven Poole, "Your Brain on Pseudoscience: The Rise of Popular Neurobollocks," *New Statesman* (blog), September 6, 2012, https://www.newstatesman.com/long-reads/2012/09/your-brain-pseudoscience-rise-popular-neurobollocks.

3. Quote from OHBM (Organization for Human Brain Mapping) interview by Peter Bandettini, "S2 EP21: Brain Wide Association Studies," *OHBM Neurosalience* podcast, Spotify for Podcasters, accessed August 14, 2022, https://anchor.fm/ohbm/episodes/S2-EP21-Brain-Wide-Association-Studies-e1juv4d.

4. Scott Marek et al., "Reproducible Brain-Wide Association Studies Require Thousands of Individuals," *Nature* 603 (March 2022): 654–660, https://doi.org/10.1038/s41586-022-04492-9.

5. John P. A. Ioannidis, "Why Most Published Research Findings Are False," *PLoS Med* 2, no. 8 (2005): e124, https://doi.org/10.1371/journal.pmed.0020124; Denes Szucs and John P. A. Ioannidis, "Empirical Assessment of Published Effect Sizes and Power in the Recent Cognitive Neuroscience and Psychology Literature," *PLoS Biology* 15, no. 3 (2017): e2000797, https://doi.org/10.1371/journal.pbio.2000797.

6. Denes Szucs and John P. A. Ioannidis, "Sample Size Evolution in Neuroimaging Research: An Evaluation of Highly-Cited Studies (1990–2012) and of Latest Practices (2017–2018) in High-Impact Journals," *NeuroImage* 221 (November 2020): 117164, https://doi.org/10.1016/j.neuroimage.2020.117164; Open Science Collaboration, "Estimating the Reproducibility of Psychological Science," *Science* 349, no. 6251 (August 28, 2015): aac4716, https://doi.org/10.1126/science.aac4716.

7. For examples, see Rick O. Gilmore et al., "Progress toward Openness, Transparency, and Reproducibility in Cognitive Neuroscience," *Annals of the New York Academy of Sciences* 1396, no. 1 (May 2017): 5–18, https://doi.org/10.1111/nyas.13325; Russell A. Poldrack et al., "Scanning the Horizon: Towards Transparent and Reproducible Neuroimaging Research," *Nature Reviews Neuroscience* 18 (2017): 115–126, https://doi.org/10.1038/nrn.2016.167; Russell A. Poldrack and Tal Yarkoni, "From Brain Maps to Cognitive Ontologies: Informatics and the Search for Mental Structure," *Annual Review of Psychology* 67 (2016): 587–612, https://doi.org/10.1146/annurev-psych-122414-033729.

8. Marek et al., "Reproducible Brain-Wide Association Studies Require Thousands of Individuals," 654–655.

9. In some cases, rather convoluted composite indices are derived and employed: Roger E. Beaty et al., "Robust Prediction of Individual Creative Ability from Brain Functional Connectivity," *Proceedings of the National Academy of Sciences* 115, no. 5 (January 16, 2018), 1087–1092, 201713532, https://doi.org/10.1073/pnas.1713532115.

Task-related brain network connectivity analyses were conducted in which correlations were carried out between "all edges in this network with individual creativity scores extracted via latent variable modeling" (1088). "The higher order creativity factor was indicated by two lower order latent factors: in-scanner ratings and laboratory-based ratings. . . . A 'creative behavior and achievement' variable was also modeled" (1091). This means that the creativity scores employed in the correlations with brain connectivity measures in this study featured the integration of (1) performance metrics recorded during the brain imaging, (2) performance metrics recorded after the brain imaging, and (3) nonimaging related responses on questionnaires metrics that assessed creativity-relevant behavior and achievement.

10. Here is an excerpt from a 2022 TED-Ed video: "Scientists think there may be two essential parts to this process: a generative phase of free-flowing ideas and spontaneous thoughts courtesy of the default mode network, followed by a process of selecting, developing, and pursuing the best ideas from that generative burst driven by logical thinking thanks to the executive network. A host of imaging studies suggest that these two networks working in sync is a crucial condition for creative thinking. Taken together, the evidence clearly suggests that the logical realm of the executive network and the imaginative realm of the default mode network are closely related." TED-Ed, "The Benefits of Daydreaming—Elizabeth Cox," posted on September 17, 2022, 2:25, https://ed.ted.com/lessons/the-benefits-of -daydreaming-elizabeth-cox.

11. For a general overview on functional neuroimaging analyses, see Bradley R. Buchbinder, "Chapter 4—Functional Magnetic Resonance Imaging," *Handbook of Clinical Neurology* 135 (2016): 61–92, https:// doi.org/10.1016/B978-0-444-53485-9.00004-0.

12. Bharat Biswal et al., "Functional Connectivity in the Motor Cortex of Resting Human Brain Using Echo-Planar MRI," *Magnetic Resonance in Medicine* 34, no. 4 (October 1995): 537, https://doi.org/10.1002 /mrm.1910340409.

13. Gordon L. Shulman et al., "Common Blood Flow Changes across Visual Tasks: I. Increases in Subcortical Structures and Cerebellum but

Not in Nonvisual Cortex," *Journal of Cognitive Neuroscience* 9, no. 5 (1997): 624–647, https://doi.org/10.1162/jocn.1997.9.5.624.

14. Gordon L. Shulman et al., "Common Blood Flow Changes across Visual Tasks: II. Decreases in Cerebral Cortex," *Journal of Cognitive Neuroscience* 9, no. 5 (1997): 648–663, https://doi.org/10.1162/jocn .1997.9.5.648.

15. Marcus E. Raichle et al., "A Default Mode of Brain Function," *Proceedings of the National Academy of Sciences of the United States of America* 98, no. 2 (January 16, 2001): 676, https://doi.org/10.1073/pnas.98 .2.676.

16. Michael D. Greicius et al., "Functional Connectivity in the Resting Brain: A Network Analysis of the Default Mode Hypothesis," *Proceedings of the National Academy of Sciences of the United States of America* 100, no. 1 (December 27, 2002): 253–258, https://doi.org/10 .1073/pnas.0135058100.

17. Marcus E. Raichle, "The Brain's Dark Energy," *Scientific American* 302, no. 3 (2010): 44–49.

18. Randy L Buckner, Jessica R Andrews-Hanna, and Daniel L Schacter, "The Brain's Default Network: Anatomy, Function, and Relevance to Disease," *Annals of the New York Academy of Sciences* 1124, no. 1 (March 2008): 1–38, https://doi.org/10.1196/annals.1440.011; Marcus E. Raichle, "The Brain's Default Mode Network," *Annual Review of Neuroscience* 38 (July 2015): 433–447, https://doi.org/10 .1146/annurev-neuro-071013-014030; Randy L. Buckner and Lauren M. DiNicola, "The Brain's Default Network: Updated Anatomy, Physiology and Evolving Insights," *Nature Reviews Neuroscience* 20, no. 10 (2019): 593–608, https://doi.org/10.1038/s41583-019-0212-7.

19. J. R. Binder et al., "Conceptual Processing during the Conscious Resting State: A Functional MRI Study," *Journal of Cognitive Neuroscience* 11, no. 1 (January 1999): 80–95, https://doi.org/10.1162/08989 2999563265.

20. D. A. Gusnard et al., "Medial Prefrontal Cortex and Self-Referential Mental Activity: Relation to a Default Mode of Brain Function," *Proceedings of the National Academy of Sciences of the United*

States of America 98, no. 7 (March 20, 2001): 4259–4264, https://doi .org/10.1073/pnas.071043098.

21. P. K. McGuire et al., "Brain Activity during Stimulus Independent Thought," *Neuroreport* 7, no. 13 (September 2, 1996): 2095–2099.

22. Malia F. Mason et al., "Wandering Minds: The Default Network and Stimulus-Independent Thought," *Science* 315, no. 5810 (January 19, 2007): 393–395, https://doi.org/10.1126/science.1131295.

23. Jessica R. Andrews-Hanna et al., "Evidence for the Default Network's Role in Spontaneous Cognition," *Journal of Neurophysiology* 104, no. 1 (July 2010): 322–335, https://doi.org/10.1152/jn.00830.2009.

24. Jessica R. Andrews-Hanna, "The Brain's Default Network and Its Adaptive Role in Internal Mentation," *The Neuroscientist* 18, no. 3 (June 2012): 251–270, https://doi.org/10.1177/1073858411403316.

25. B. T. Thomas Yeo et al., "The Organization of the Human Cerebral Cortex Estimated by Intrinsic Functional Connectivity," *Journal of Neurophysiology* 106, no. 3 (September 2011): 1125–1265, https:// doi.org/10.1152/jn.00338.2011.

26. Hikaru Takeuchi et al., "Failing to Deactivate: The Association between Brain Activity during a Working Memory Task and Creativity," *NeuroImage* 55, no. 2 (March 15, 2011): 681, https://doi.org /10.1016/j.neuroimage.2010.11.052.

27. The dominant view at the time was that the functional role of the precuneus is in self-consciousness, particularly in relation to episodic retrieval and "self-centered spatial imagery": Andrea E. Cavanna and Michael R. Trimble, "The Precuneus: A Review of Its Functional Anatomy and Behavioural Correlates," *Brain: A Journal of Neurology* 129, no. 3 (March 2006): 564–583, https://doi.org/10.1093/brain/awl004. See also Michael F. Land, "Do We Have an Internal Model of the Outside World?," *Philosophical Transactions of the Royal Society B* 369, no. 1636 (2014): 20130045, https://doi.org/10.1098/rstb.2013.0045.

28. Nancy C. Andreasen, "A Journey into Chaos: Creativity and the Unconscious," *Mens Sana Monographs* 9, no. 1 (December–January 2011): 50, https://www.ncbi.nlm.nih.gov/pmc/articles/PMC3115302/.

29. Ellamil et al., "Evaluative and Generative Modes of Thought during the Creative Process," 1783. Notably, as there was no resting baseline included in the study design, it is impossible to verify if the activity patterns seen reflect actual "activations" associated with creative idea generation and evaluation over and above rest.

30. Charles J. Limb and Allen R. Braun, "Neural Substrates of Spontaneous Musical Performance: An FMRI Study of Jazz Improvisation," *PloS ONE* 3, no. 2 (2008): 3, 4, https://doi.org/10.1371/journal .pone.0001679. This is a skewed interpretation given that the findings show areas known to be involved in self-referential cognition (dorsal medial prefrontal cortex) were also deactivated in this study; see Taylor W. Schmitz and Sterling C. Johnson, "Relevance to Self: A Brief Review and Framework of Neural Systems Underlying Appraisal," *Neuroscience & Biobehavioral Reviews* 31, no. 4 (2007): 585–596, https://doi.org/10.1016/j.neubiorev.2006.12.003. Note also that Limb and Braun did not lean on network-level explanations when interpreting their findings, but their work is frequently cited by others who do.

31. Free associative idea generation during creative thinking is attributed to the workings of the precuneus by Takeuchi et al., "Failing to Deactivate"; the dorsolateral prefrontal cortex by Limb and Braun, "Neural Substrates of Spontaneous Musical Performance"; the MTL network by Ellamil et al., "Evaluative and Generative Modes of Thought during the Creative Process"; and the default mode network by Andreasen, "A Journey into Chaos."

32. A case in point is Limb and Braun, "Neural Substrates of Spontaneous Musical Performance," in which the distinction between the dorsal medial prefrontal cortex (a default mode network region that was deactivated during improvisation) and the frontopolar medial prefrontal cortex (a default mode network region that was activated during improvisation) was unintentionally blurred. In one section of the paper, it is specified that "within the prefrontal cortex, a dissociated pattern of activity was seen during improvisation. This was characterized by widespread *deactivation* that included almost all of the lateral prefrontal cortices . . . as well as *dorsal portions of the medial*

prefrontal cortex (MPFC). However, this broad pattern of deactivation was also accompanied by focal *activation of the frontal polar portion of the MPFC*" (3; italics mine). However, figure 3 merely states that improvisation was associated with "*medial prefrontal cortex activation*, dorsolateral prefrontal cortex deactivation, and sensorimotor activation" (italics mine). This paper has been cited more than nine hundred times to date. And a large proportion of these citing articles focus on the assertions from figure 3 (and not the abstract or the rest of the article) as representing the overall findings from this paper.

33. Here are two examples of the misattribution problem in relation to the aforementioned study (see note 32). The first is from a paper by the very same research group: "The most striking feature of lyrical improvisation detected by direct comparison of freestyle and conventional performance, in large part consistent with the previous study of melodic improvisation [Limb and Braun, "Neural Substrates of Spontaneous Musical Performance"], was a dissociated pattern of activity within the prefrontal cortex: increases in activity throughout the MPFC, extending from the frontal pole to the border of the pre-SMA, and simultaneous decreases in the DLPFC, from its orbital to superior regions": Siyuan Liu et al., "Neural Correlates of Lyrical Improvisation: An fMRI Study of Freestyle Rap," *Scientific Reports* 2, no. 834 (2012): 4, https://doi.org/10.1038/srep00834. The second is from another research group who stated: "The first study to implicate DMN in improvisation used a functional Magnetic Resonance Imaging (fMRI) paradigm where expert jazz pianists switched between playing prelearned and improvised material [Limb and Braun, "Neural Substrates of Spontaneous Musical Performance"]. Compared to playing learned sequences, improvisation showed increased activation of DMN regions and concurrent deactivation of ECN regions": Victor M. Vergara et al., "Functional Network Connectivity during Jazz Improvisation," *Scientific Reports* 11, no. 19036 (2021): 2, https://doi.org/10.1038/s41598-021-98332-x.

34. Rex E. Jung et al., "The Structure of Creative Cognition in the Human Brain," *Frontiers in Human Neuroscience* 7 (July 8, 2013), https://doi.org/10.3389/fnhum.2013.00330.

35. Leh Woon Mok, "The Interplay between Spontaneous and Controlled Processing in Creative Cognition," *Frontiers in Human Neuroscience* 8 (2014): 663, https://doi.org/10.3389/fnhum.2014.00663.

36. Like that of Liane Gabora who noted that creativity involves both divergent and convergent thinking and as well as "the capacity to spontaneously shift between them according to the situation, referred to as contextual focus": Liane Gabora, "Revenge of the 'Neurds': Characterizing Creative Thought in Terms of the Structure and Dynamics of Memory," *Creativity Research Journal* 22, no. 1 (2010): 10, https://doi.org/10.1080/10400410903579494.

37. Roger E. Beaty et al., "Creative Cognition and Brain Network Dynamics," *Trends in Cognitive Sciences* 20, no. 2 (February 2016): 87–88, https://doi.org/10.1016/j.tics.2015.10.004.

38. Here is a selection of such studies with the type of neuroimaging analysis employed indicated in square brackets: [fMRI: Resting state brain connectivity analysis] Roger E. Beaty et al., "Brain Networks of the Imaginative Mind: Dynamic Functional Connectivity of Default and Cognitive Control Networks Relates to Openness to Experience," *Human Brain Mapping* 39 (2018): 811–821, https://doi.org/10.1002/hbm.23884; [fMRI MVPA: Multivariate pattern analysis] James Lloyd-Cox, Qunlin Chen, and Roger E. Beaty, "The Time Course of Creativity: Multivariate Classification of Default and Executive Network Contributions to Creative Cognition over Time," *Cortex* 156 (November 2022): 90–105, https://doi.org/10.1016/j.cortex.2022.08.008; [fMRI: Resting-state brain regions-of-interest analysis] Jingyi Zhang et al., "Retrieval Flexibility Links to Creativity: Evidence from Computational Linguistic Measure," *Cerebral Cortex* 33, no. 8 (April 2023), 4964–4976, https://doi.org/10.1093/cercor/bhac392; [fMRI: Task-related whole-brain functional connectivity analysis] Hua Xie et al., "Spontaneous and Deliberate Modes of Creativity: Multitask Eigen-Connectivity Analysis Captures Latent Cognitive Modes during Creative Thinking," *NeuroImage* 243 (November 2021): 118531, https://doi.org/10.1016/j.neuroimage.2021.118531; [fMRI: Task-related regions-of-interest analysis] Evangelia G. Chrysikou et al., "Differences in Brain Activity Patterns during Creative Idea Generation between Eminent and Non-eminent Thinkers," *NeuroImage* 220

(October 15, 2020): 117011, https://doi.org/10.1016/j.neuroimage
.2020.117011; [fMRI: Task-related regions-of-interest connectivity
analysis] Naama Mayseless, Ayelet Eran, and Simone G. Shamay-
Tsoory, "Generating Original Ideas: The Neural Underpinning of
Originality," *NeuroImage* 116 (August 1, 2015): 232–239, https://doi
.org/10.1016/j.neuroimage.2015.05.030; [MRI: Voxel-based lesion-
deficit and disconnection-deficit mapping] David Bendetowicz et al.,
"Two Critical Brain Networks for Generation and Combination of
Remote Associations," *Brain* 141, no. 1 (2018): 217–233, https://doi
.org/10.1093/brain/awx294.

39. "This analysis revealed a significant relationship between activity to divergent thinking and the semantic control network": Lucy
S. Cogdell-Brooke et al., "A Meta-analysis of Functional Magnetic
Resonance Imaging Studies of Divergent Thinking Using Activation
Likelihood Estimation," *Human Brain Mapping* 41, no. 17 (December 2020): 5073, https://doi.org/10.1002/hbm.25170; "To a certain
extent, these brain regions may be separated into two brain networks,
namely, semantic (such as left IFG, left MFG, left IPL, and left MTG)
and cognitive control systems (right ACC, right MFG, and right IPL)":
Xin Wu et al., "A Meta-analysis of Neuroimaging Studies on Divergent Thinking Using Activation Likelihood Estimation," *Human Brain
Mapping*, April 18, 2015, 2715, https://doi.org/10.1002/hbm.22801.

40. For a contemporary theoretical account that outlines the dynamics between the semantic and the central executive networks in
creative cognition, see Anna Abraham, "Creative Thinking as Orchestrated by Semantic Processing vs. Cognitive Control Brain Networks,"
Frontiers in Human Neuroscience 8 (2014), https://doi.org/10.3389
/fnhum.2014.00095. Classic accounts are models that emphasize the
role of knowledge storage and knowledge access: S. A. Mednick, "The
Associative Basis of the Creative Process," *Psychological Review* 69 (May
1962): 220–232; Gerald A. Mendelsohn, "Associative and Attentional
Processes in Creative Performance," *Journal of Personality* 44 (1974):
341–369.

41. Episodic memory-specific inductions have been shown to accompany improvements in fluency (or the total number of responses
generated), but not in the originality (or uniqueness of generated

responses) during the alternate uses task: Kevin P. Madore, Donna Rose Addis, and Daniel L. Schacter, "Creativity and Memory: Effects of an Episodic-Specificity Induction on Divergent Thinking," *Psychological Science* 26, no. 9 (2015): 1461–1468, https://doi.org/10.1177/0956797615591863. Conversely, semantic inductions lead to improvements in originality: Roger E. Beaty et al., "Default Network Contributions to Episodic and Semantic Processing during Divergent Creative Thinking: A Representational Similarity Analysis," *NeuroImage* 209 (April 2020): 116499, https://doi.org/10.1016/j.neuroimage.2019.116499. This fits with behavioral studies that show that episodic strategies are common early in the ideation process when familiar (and thereby less original) responses are readily elicited, whereas semantic strategies are applied later in the ideation process and are tied to the generation of more novel ideas: Kenneth J. Gilhooly et al., "Divergent Thinking: Strategies and Executive Involvement in Generating Novel Uses for Familiar Objects," *British Journal of Psychology* 98, no. 4 (November 2007): 611–625, https://doi.org/10.1111/j.2044-8295.2007.tb00467.x. Indeed, direct cortical stimulation of default mode brain regions was shown to decrease the fluency of generated responses, but not their originality: Ben Shofty et al., "The Default Network Is Causally Linked to Creative Thinking," *Molecular Psychiatry* 27 (2022): 1848–1854, https://doi.org/10.1038/s41380-021-01403-8.

42. Daniel C. Zeitlen et al., "The Creative Mind in Daily Life: How Cognitive and Affective Experiences Relate to Creative Thinking and Behavior," *Psychology of Aesthetics, Creativity, and the Arts* (2022), https://doi.org/10.1037/aca0000537.

43. For an example of a time-course study in the neuroscience of creativity, see Lloyd-Cox, Chen, and Beaty, "The Time Course of Creativity."

44. Even when designing clever experimental paradigms that allow for the dissociation of these phases from a cognitive-behavioral standpoint, there are added considerations to reckon with from an fMRI design standpoint. See Gary H. Glover, "Overview of Functional Magnetic Resonance Imaging," *Neurosurgery Clinics of North America* 22, no. 2 (April 2011): 133–139, https://doi.org/10.1016/j.nec.2010

.11.001. At the very minimum, there needs to be a temporal "jitter" between the trial events to be able to make a case for phase-based interpretations.

45. "Neuromodulatory effects, such as those effected by arousal, attention, memory, and so on, are slow and have reduced spatiotemporal resolution and specificity. . . . The fMRI signal cannot easily differentiate between function-specific processing and neuromodulation, between bottom-up and top-down signals, and it may potentially confuse excitation and inhibition. The magnitude of the fMRI signal cannot be quantified to reflect accurately differences between brain regions, or between tasks within the same region": Nikos K. Logothetis, "What We Can Do and What We Cannot Do with fMRI," *Nature* 453 (June 2008): 871, 877, https://doi.org/10.1038/nature 06976.

46. For guidance on sound practices to adopt when carrying out BWAS using smaller datasets, see Monica D. Rosenberg and Emily S. Finn, "How to Establish Robust Brain–Behavior Relationships without Thousands of Individuals," *Nature Neuroscience* 25 (July 2022): 835–837, https://doi.org/10.1038/s41593-022-01110-9; Peter A. Bandettini et al., "The Challenge of BWAs: Unknown Unknowns in Feature Space and Variance," *Med* 3, no. 8 (August 12, 2022): 526–531, https://doi.org/10.1016/j.medj.2022.07.002.

47. Jeffrey R. Binder and Rutvik H. Desai, "The Neurobiology of Semantic Memory," *Trends in Cognitive Sciences* 15, no. 11 (November 2011): 527–536, https://doi.org/10.1016/j.tics.2011.10.001.

48. William W. Seeley, "The Salience Network: A Neural System for Perceiving and Responding to Homeostatic Demands," *Journal of Neuroscience* 39, no. 50 (December 11, 2019): 9878–9882, https://doi.org/10.1523/JNEUROSCI.1138-17.2019; Weidong Cai et al., "Dynamic Causal Brain Circuits during Working Memory and Their Functional Controllability," *Nature Communications* 12 (June 2021): 3314, https://doi.org/10.1038/s41467-021-23509-x.

49. When discussing Ellamil et al., "Evaluative and Generative Modes of Thought during the Creative Process," the authors note that "idea generation was associated with widespread activity of default regions,

whereas idea evaluation was associated with both default (e.g., MPFC and PCC) and control network activity (e.g., DLPFC and ACC)": Beaty et al., "Creative Cognition and Brain Network Dynamics," 92.

50. This problem of skewed interpretations even occurs with findings from meta-analyses. Take the case of this recent paper: Xiaofei Wu et al., "Neural Correlates of Novelty and Appropriateness Processing in Cognitive Reappraisal," *Biological Psychology* 170 (April 2022): 2, https://doi.org/10.1016/j.biopsycho.2022.108318 where the authors state: "Two other meta-analyses revealed that the inferior parietal lobe (IPL) and medial PFC (mPFC) (both belong to the DMN) . . . were activated in creative tasks [Cogdell-Brooke et al., "A Meta-analysis of Functional Magnetic Resonance Imaging Studies"; Wu et al., "Verbal Creativity in Semantic Variant Primary Progressive Aphasia"]." However, the first of these papers in actuality says something quite different about the inferior parietal lobe, namely that it belongs to another brain network: "Activation Likelihood Estimation was then used to integrate the neuroimaging results across studies. This revealed four clusters: the left inferior parietal lobe; the left inferior frontal and precentral gyrus; the superior and medial frontal gyrus and the right cerebellum. These regions are key in the semantic network, important for flexible retrieval of stored knowledge, highlighting the role of this network in divergent thinking": Cogdell-Brooke et al., "A Meta-analysis of Functional Magnetic Resonance Imaging Studies," 5057. So even though the authors of the cited meta-analysis conclude that it is the semantic network that is vital in divergent thinking, contemporary literature citation practices in neuroscience are such that any other researcher can choose to ignore the conclusions of the original authors and present their own take instead without even acknowledging the disparity in interpretation. This only happens because most individual brain regions (particularly association cortices) are tied to multiple networks.

51. Randy L. Buckner, Fenna M. Krienen, and B. T. Thomas Yeo, "Opportunities and Limitations of Intrinsic Functional Connectivity MRI," *Nature Neuroscience* 16 (July 2013): 832–837, https://doi.org/10.1038/nn.3423.

52. Stephen M. Smith et al., "Correspondence of the Brain's Functional Architecture during Activation and Rest," *Proceedings of the National Academy of Sciences* 106, no. 31 (August 4, 2009): 13040–13045, https://doi.org/10.1073/pnas.0905267106; W. Dale Stevens and R. Nathan Spreng, "Resting-State Functional Connectivity MRI Reveals Active Processes Central to Cognition: Cognition and Resting-State Functional Connectivity," *Wiley Interdisciplinary Reviews: Cognitive Science* 5, no. 2 (March/April 2014): 233–245, https://doi.org /10.1002/wcs.1275.

53. Simon W. Davis et al., "Resting-State Networks Do Not Determine Cognitive Function Networks: A Commentary on Campbell and Schacter (2016)," *Language, Cognition and Neuroscience* 32, no. 6 (2017): 669, https://doi.org/10.1080/23273798.2016.1252847.

54. Emily S. Finn, "Is It Time to Put Rest to Rest?," *Trends in Cognitive Sciences* 25, no. 12 (December 2021): 1021–1032, https://doi.org/10 .1016/j.tics.2021.09.005; Alexa M. Morcom and Paul C. Fletcher, "Does the Brain Have a Baseline? Why We Should Be Resisting a Rest," *NeuroImage* 37, no. 4 (October 2007): 1073–1082, https://doi .org/10.1016/j.neuroimage.2006.09.013.

55. Kevin Murphy, Rasmus M. Birn, and Peter A. Bandettini, "Resting-State fMRI Confounds and Cleanup," *NeuroImage*, Mapping the Connectome, 80 (October 15, 2013): 349–359, https://doi.org/10.1016/j .neuroimage.2013.04.001.

56. Sanam Maknojia et al., "Resting State fMRI: Going Through the Motions," *Frontiers in Neuroscience* 13 (2019), https://www.frontiersin .org/articles/10.3389/fnins.2019.00825.

57. Toshikazu Kawagoe, Keiichi Onoda, and Shuhei Yamaguchi, "Different Pre-Scanning Instructions Induce Distinct Psychological and Resting Brain States during Functional Magnetic Resonance Imaging," *European Journal of Neuroscience* 47, no. 1 (2018): 77–82, https://doi .org/10.1111/ejn.13787; Christopher Benjamin et al., "The Influence of Rest Period Instructions on the Default Mode Network," *Frontiers in Human Neuroscience* 4 (December 2010): 218, https://doi.org/10.3389 /fnhum.2010.00218.

58. Javier Gonzalez-Castillo et al., "How to Interpret Resting-State fMRI: Ask Your Participants," *Journal of Neuroscience* 41, no. 6 (February 10, 2021): 1130–1141, https://doi.org/10.1523/JNEUROSCI.1786 -20.2020.

59. Finn, "Is It Time to Put Rest to Rest?," 1030.

60. Owing to methodological limitations, caution is advised and making strong interpretative claims based on the sign of correlations between networks is not encouraged: Buckner, Krienen, and Yeo, "Opportunities and Limitations of Intrinsic Functional Connectivity MRI."

61. R. Nathan Spreng et al., "Default Network Activity, Coupled with the Frontoparietal Control Network, Supports Goal-Directed Cognition," *NeuroImage* 53, no. 1 (October 15, 2010): 303–317, https://doi .org/10.1016/j.neuroimage.2010.06.016.

62. Kieran C. R. Fox et al., "The Wandering Brain: Meta-analysis of Functional Neuroimaging Studies of Mind-Wandering and Related Spontaneous Thought Processes," *NeuroImage* 111 (May 2015), 611– 621, https://doi.org/10.1016/j.neuroimage.2015.02.039.

63. Ben M. Crittenden, Daniel J. Mitchell, and John Duncan, "Recruitment of the Default Mode Network during a Demanding Act of Executive Control," *eLife* 4 (2015): 9, https://doi.org/10.7554 /eLife.06481.

64. Sam J. Gilbert et al., "Comment on 'Wandering Minds: The Default Network and Stimulus-Independent Thought,'" *Science* 317, no. 5834 (July 6, 2007): 43, https://doi.org/10.1126/science.1140801.

65. Darya L. Zabelina and Jessica R. Andrews-Hanna, "Dynamic Network Interactions Supporting Internally-Oriented Cognition," *Current Opinion in Neurobiology* 40 (October 2016): 86–93, https://doi .org/10.1016/j.conb.2016.06.014.

66. Lorenzo Mancuso et al., "Tasks Activating the Default Mode Network Map Multiple Functional Systems," *Brain Structure & Function* 227 (June 2022): 1711–1734, https://doi.org/10.1007/s00429-022 -02467-0.

67. Yaara Yeshurun, Mai Nguyen, and Uri Hasson, "The Default Mode Network: Where the Idiosyncratic Self Meets the Shared Social World," *Nature Reviews Neuroscience* 22 (March 2021): 190, https://doi .org/10.1038/s41583-020-00420-w. For a parallel view of the default mode network as orchestrating long-term mental states (including personal and vigilance-based processes), see Colin Klein, "The Brain at Rest: What It Is Doing and Why That Matters," *Philosophy of Science* 81, no. 5 (December 2014): 974–985, https://doi.org/10.1086/677692.

68. Anna Abraham, "The Imaginative Mind," *Human Brain Mapping* 37, no. 11 (November 2016): 4203, https://doi.org/10.1002/hbm .23300. See also Anna Abraham, "Surveying the Imagination Landscape," in *The Cambridge Handbook of the Imagination*, ed. Anna Abraham (Cambridge: Cambridge University Press, 2020), 1–10, https:// doi.org/10.1017/9781108580298.001.

69. Paul T. Sowden, Andrew Pringle, and Liane Gabora, "The Shifting Sands of Creative Thinking: Connections to Dual-Process Theory," *Thinking & Reasoning* 21, no. 1 (2015): 40–60, https://doi.org/10.1080 /13546783.2014.885464; Anna Abraham, "The Forest versus the Trees: Creativity, Cognition and Imagination," in *Cambridge Handbook of the Neuroscience of Creativity*, ed. Rex E. Jung and Oshin Vartanian (New York: Cambridge University Press, 2018), 195–210.

70. For some prominent examples, see Arne Dietrich, "Who's Afraid of a Cognitive Neuroscience of Creativity?," *Methods* (San Diego, CA) 42, no. 1 (May 2007): 22–27, https://doi.org/10.1016/j.ymeth.2006 .12.009; Dietrich and Kanso, "A Review of EEG, ERP, and Neuroimaging Studies"; Dietrich, *How Creativity Happens in the Brain*; Rosalind Arden et al., "Neuroimaging Creativity: A Psychometric View," *Behavioural Brain Research* 214, no. 2 (December 25, 2010): 143–156, https:// doi.org/10.1016/j.bbr.2010.05.015; Keith Sawyer, "The Cognitive Neuroscience of Creativity: A Critical Review," *Creativity Research Journal* 23, no. 2 (2011): 137–154, https://doi.org/10.1080/10400419.2011 .571191; Abraham, "The Promises and Perils of the Neuroscience of Creativity"; Abraham, *The Neuroscience of Creativity*.

71. Kris, *Psychoanalytic Explorations in Art*.

72. Freud (1920/1953) as cited in Suler, "Primary Process Thinking and Creativity," 146.

73. Sigmund Freud, *Introductory Lectures on Psycho-Analysis*, trans. Joan Riviere (London: George Allen & Unwin Ltd, 1922), 314–315.

Afterword

1. Lev Semenovich Vygotsky, *The Collected Works of L. S. Vygotsky, Vol. 3: Problems of the Theory and History of Psychology: Including the Chapter on the Crisis of Psychology*, ed. Robert W. Rieber and Jeffrey L. Wollock, trans. René van der Veer (New York: Plenum Press, 1997), 280.

2. Examples include astrology and dream interpretation, fields that seek to provide explanatory mechanisms for understanding ourselves and what the future holds for us: Maarten Boudry, Stefaan Blancke, and Massimo Pigliucci, "What Makes Weird Beliefs Thrive? The Epidemiology of Pseudoscience," *Philosophical Psychology* 28, no. 8 (2015): 1177–1198, https://doi.org/10.1080/09515089.2014.971946. Even though it is well known that there is a lack of scientific evidence for the explanations provided by such fields, it doesn't seem to matter to a sizeable portion of the population. Recent estimates indicate that over a quarter of Americans believe in astrology, for instance: Taylor Orth, "One in Four Americans Say They Believe in Astrology | YouGov," YouGovAmerica, April 26, 2022, https://today.yougov.com/topics/entertainment/articles-reports/2022/04/26/one-four-americans-say-they-believe-astrology.

3. A great example comes from the enormously influential research on mirror neurons, which were first discovered to play a key role in action recognition. Over time, several other functions were ascribed to mirror neurons (following the same principles), including action understanding, imitation, communication, and empathy. For a scathing critique of this field, see Gregory Hickok, *The Myth of Mirror Neurons: The Real Neuroscience of Communication and Cognition* (New York: W. W. Norton & Company, 2014).

4. Bronwyn T. Williams, "Never Let the Truth Stand in the Way of a Good Story: A Work for Three Voices," *College English* 65, no. 3 (2003): 290–304, https://doi.org/10.2307/3594259.

5. Brian Boyd, "The Evolution of Stories: From Mimesis to Language, from Fact to Fiction," *WIREs Cognitive Science* 9, no. 1 (January/February 2018), https://doi.org/10.1002/wcs.1444.

6. This "need for narrative to structure events" is a cognitive bias that is attributed to be one of the causes of the replication crisis in Psychology: Dorothy Bishop, "Fixing the Replication Crisis: The Need to Understand Human Psychology," *APS Observer* 32 (November 25, 2019), https://www.psychologicalscience.org/observer/fixing-the -replication-crisis-the-need-to-understand-human-psychology.

7. Deena Skolnick Weisberg et al., "The Seductive Allure of Neuroscience Explanations," *Journal of Cognitive Neuroscience* 20, no. 3 (March 2008): 470–747, https://doi.org/10.1162/jocn.2008.20040.

8. Nicholas Scurich and Adam Shniderman, "The Selective Allure of Neuroscientific Explanations," *PLoS ONE* 9, no. 9 (September 10, 2014), https://doi.org/10.1371/journal.pone.0107529.

9. Martha J. Farah and Cayce J. Hook, "The Seductive Allure of 'Seductive Allure,'" *Perspectives on Psychological Science* 8, no. 1 (January 2013): 88–90, https://doi.org/10.1177/1745691612469035; see also Neuroskeptic, "Critiquing a Classic: 'The Seductive Allure of Neuroscience Explanations,'" *Neuroskeptic* (blog), January 6, 2009, http:// neuroskeptic.blogspot.com/2009/01/critiquing-classic-seductive -allure-of.html.

10. Raymond S. Nickerson, "Confirmation Bias: A Ubiquitous Phenomenon in Many Guises," *Review of General Psychology* 2, no. 2 (1998): 175–220, https://doi.org/10.1037/1089-2680.2.2.175; Barbara Koslowski, "Scientific Reasoning: Explanation, Confirmation Bias, and Scientific Practice," in *Handbook of the Psychology of Science* (New York: Springer Publishing Company, 2013), 151–192.

11. Peer review is the quality control mechanism that was established a little over half a century ago for estimating the worthiness of academic publications: Adam Mastroianni, "The Rise and Fall of Peer Review," Substack newsletter, *Experimental History* (blog), December 13, 2022, https://experimentalhistory.substack.com/p/the -rise-and-fall-of-peer-review.

12. Jürgen Huber et al., "Nobel and Novice: Author Prominence Affects Peer Review," *Proceedings of the National Academy of Sciences* 119, no. 41 (October 11, 2022): e2205779119, https://doi.org/10.1073/pnas.2205779119.

13. Monya Baker, "1,500 Scientists Lift the Lid on Reproducibility," *Nature* 533, no. 7604 (May 1, 2016): 452–454, https://doi.org/10.1038/533452a; Ed Yong, "Psychology's Replication Crisis Is Running Out of Excuses," *The Atlantic*, November 19, 2018, sec. Science, https://www.theatlantic.com/science/archive/2018/11/psychologys-replication-crisis-real/576223/.

14. Some comprehensive resources include Dorothy V. M. Bishop, "The Psychology of Experimental Psychologists: Overcoming Cognitive Constraints to Improve Research: The 47th Sir Frederic Bartlett Lecture," *Quarterly Journal of Experimental Psychology* 73, no. 1 (January 2020): 1–19, https://doi.org/10.1177/1747021819886519; and David Randall and Christopher Welser, *The Irreproducibility Crisis of Modern Science: Causes, Consequences, and the Road to Reform, National Association of Scholars* (National Association of Scholars, 2018), https://eric.ed.gov/?id=ED600638.

15. Ben Goldacre, *I Think You'll Find It's a Bit More Complicated Than That* (London: Fourth Estate, 2015); Adam Grant, *Think Again: The Power of Knowing What You Don't Know* (New York: Viking, 2021).

16. Janet Metcalfe and Arthur P. Shimamura, eds., *Metacognition: Knowing about Knowing* (Cambridge, MA: MIT Press, 1996). The power of metacognitive self-reflection has been emphasized throughout history in diverse religious and philosophical traditions.

17. This question is a paraphrase of an assertion made by the economist, author, and academic Glenn Loury in a 2021 interview: Bari Weiss, "Wrongthink on Race with Glenn C. Loury," *Honestly with Bari Weiss* podcast, accessed September 29, 2021, https://podcasts.apple.com/us/podcast/wrongthink-on-race-with-glenn-c-loury/id1570872415?i=1000536938762: "There's the poetic truth, and there's the larger truth."

Name Index

Peterson, Jordan B., 224n40,
    228n71
Pietikainen, Petteri, 40, 172n12
Pigliucci, Massimo, 246n2
Plato, 171–172n6
Plucker, Jonathan A., 214n48
Poincaré, Henri, 186n16
Poldrack, Russell A., 232n7
Porter, Roy, 172n12
Porter, Theodore, 105
Post, Felix, 188n28
Preti, Antonio, 187n23
Prochazkova, Luisa, 189n31

Radel, Rémi, 205n68
Raichle, Marcus, 133–134
Randall, David, 248n14
Ribot, Theodule Armand, 104
Rice, Marianne, 203n55
Richards, Ruth, 177–178n50
Rilke, Rainer Maria, 25
Rivers, Anissa, 187n22
Robinson, Andrew, 104, 172n8
Robinson, Siobhan, 191n41
Rosenberg, Monica D., 241n46
Rotenberg, Vadim S., 169n86
Roth, Ilona, 203n57
Rothenberg, Albert, 41
Russ, Suzanne L., 184–185n8
Rybakowski, Janusz K.,
    178–179n55

Sacks, Oliver, 80, 217n1
Sahakian, Barbara J., 195n68
Samorini, Giorgio, 184n5
Sample, Ian, 184n7
Saporta, Arielle, 224n40

Savage, Matt, 199n32
Savulich, George, 195n68
Sawyer, Keith, 245n70
Schacter, Daniel L., 239–240n41
Schlesinger, Judith, 158n9,
    177n49
Schmid, Yasmin, 185n9
Schmidt, Timo T., 48
Schuldberg, David, 178n54
Schultz, Wolfram, 221n24
Schürhoff, Franck, 178–179n55
Schwartz, Sophie, 166n63
Seger, C. A., 169n86
Shamay-Tsoory, Simone G.,
    238–239n38
Shebalin, Vissarion, 78
Shockley, William, 93
Shofty, Ben, 239–240n41
Silvia, Paul J., 215n52
Simon, Théodore, 88
Smith, Frederick M., 171n5
Snyder, Solomon H., 218n8
Sobiecki, Jean-Francois, 185n11
Soulières, Isabelle, 198–199n25
Spearman, Charles, 86–87
Sperry, Roger, 8, 10
Stein, Morris I., 80, 158–159n12,
    207–208n84
Stendhal, 76
Suler, John R., 165n59
Szasz, Thomas Stephen,
    173–174n23
Szigeti, Balázs, 190n35
Szumilas, Magdalena, 176n42

Takeuchi, Hikaru, 230n83,
    236n31

Tammet, Daniel, 199n32
Tan, Jacinth J. X., 203n55
Taylor, Calvin W., 212n27
ten Berge, Jos, 187–188n24
Terman, Lewis, 90–93, 95, 103
Thaker, Gunvant K., 175–176n38
Thorndike, Edward, 104
Thorndike, Robert, 97
Thurstone, Louis, 94
Tobin, Lucy, 197n10
Tolstoy, Lev, 173n20
Torrance, E. Paul, 212n27,
    214n44
Torrico, Tyler J., 191n39
Treffert, Darold A., 69, 198n23,
    207n78
Trimble, Michael R., 235n27
Turgenev, Ivan, 76
Turner, Casey E., 166n63

Vellante, Marcello, 187n23
Vergara, Victor M., 237n33
Vernon, P. E., 213n38
Vogel, Philip, 10
Vygotsky, Lev Semenovich, 63,
    64, 151, 216n59

Wallach, Michael, 95–96,
    212n31
Wallas, Graham, 186n16
Waterhouse, Lynn, 209n11
Weisberg, Robert W., 158n9
Welser, Christopher, 248n14
Wernicke, Carl, 4–5
Whitehead, Alfred North, xiv,
    159n13
Whitehead, James, 182n77

Wieder, Charles G., 170n91
Wigan, Arthur, 4
Williamson, Rachel L., 216n60
Wiltshire, Stephen, 199n32
Windmann, Sabine, 178n54
Wolff, Ulrika, 202n53
Wu, Xiaofei, 242n50

Xin Wu, 239n39

Yaden, David B., 195n68
Yaden, Mary E., 195n68
Yalincetin, Berna, 175–176n38
Yarkoni, Tal, 232n7
Yeo, B. T. Thomas, 244n60
Yuen-Siang Ang, 225n47

Zabelina, Darya, 224n40
Zaidel, Dahlia, 79, 201n48,
    207n79
Zhang, Jingyi, 238–239n38

Subject Index

Abductive reasoning, 86
Action, 5–6, 47, 48, 68, 73, 110,
    113
ADHD (attention-deficit
    hyperactivity disorder), 72,
    75, 115–116, 224n42
Affect, 30, 112, 121–122, 138
Afterglow state (hallucinogenic
    experience), 46, 56, 57, 58
Aha moment, 17. *See also* Insight
Alzheimer's disease, 79
Antipsychiatry movement,
    173–174n23
Aphasia, 76, 78
  left hemisphere, 5
  motor, 4
  primary progressive (PPA), 71
  sensory, 5
Appositional capacity, 11–12
Arcuate fasciculus, 160n17
Artificial intelligence, 84
Artistic style, 55, 72, 76–77, 79,
    201n48
Artists
  atypical brains, 75–81
  cognitive biases, 37

drug use, 49, 40
madness, 24
mental illness, 25, 31, 41
psychedelic experience, 55
risky career paths, 38–39
schizotypal traits, 34, 35
Art therapy, 27–28
Asymmetries. *See also* Dualistic
    models
  experimentally derived,
    17–18
  philosophically motivated,
    17–18
Attention, 10, 30, 42, 69, 76,
    113, 116, 123, 126, 133,
    136, 139
  "leaky" and "flexible," 224n40
Atypicality and atypical brains,
    63–67
  acquired savant syndrome,
    70–72
  ADHD (attention-deficit
    hyperactivity disorder), 72,
    75, 115–116
  autism spectrum disorder, 67,
    69, 70, 72–73, 74